Praise for *The Politically Incorrect Guide*™ *to Global Warming and Environmentalism*

"Future generations will wonder in bemused amazement that the early twenty-first century's developed world went into hysterical panic over a globally averaged temperature increase of a few tenths of a degree, and, on the basis of gross exaggerations of highly uncertain computer projections combined into implausible chains of inference, proceeded to contemplate a roll-back of the industrial age.

"Chris Horner takes us through this story, including fascinating investigations into the role of major industries and the muzzling of science, with verve, humor, and a genuine concern for accuracy. His descriptions will provoke laughter as well as recognition that tears are more in order."

> —**Dr. Richard S. Lindzen,** Alfred P. Sloan Professor of Atmospheric Sciences, MIT; member of the National Academy of Sciences; and former lead author, UN Intergovernmental Panel on Climate Change

"*The Politically Incorrect Guide*™ *to Global Warming* combines the scientific truths and a sobering perspective that Hollywood, the media, and environmental extremists don't want you to know. Chris Horner's book should be required reading by every high school and university student in America. Finally, someone has written a definitive resource to debunk global warming alarmism."

> —**Senator James Inhofe** (R-Okla.), former chairman of the Environment & Public Works Committee

"In *The Politically Incorrect Guide*™ *to Global Warming and Environmentalism*, Chris Horner exposes the myths about the most topical topic of today—and tomorrow. He reveals the inconvenient truths raised by the skeptics and exposes the convenient lies used by warmoholics. This accessible book will arm you to cut through the hot air in this ongoing national debate. Highly recommended also as source material for letter-writers to congressmen and senators— and to newspapers. A must for high school and college debaters."

> —**Dr. Fred Singer**, former director of the U.S. Weather Satellite Service; professor emeritus of environmental sciences, University of Virginia

"Chris Horner has been a staunch champion for common sense in the face of mounting media hysteria about the climate. *The Politically Incorrect Guide*™ *to Global Warming* takes a hard-headed, realistic look at both the science and the politics of the issue. It will be an invaluable handbook for all of us who want to leave our grandchildren not only a sustainable world, but also a viable economy."

> —**Roger Helmer**, member of European Parliament; American Legislative Exchange Council 2006 International Legislator of the Year

The **Politically Incorrect Guide**™ to

GLOBAL WARMING
and Environmentalism

The **Politically Incorrect Guide**™ to

GLOBAL WARMING
and Environmentalism

✷ ✷ ✷ ✷ ✷ ✷

CHRISTOPHER C. HORNER

Since 1947
REGNERY
PUBLISHING, INC.
An Eagle Publishing Company • Washington, DC

Cataloging-in-Publication data on file with the Library of Congress

ISBN 978-1-59698-501-8

Published in the United States by

Regnery Publishing, Inc.

One Massachusetts Avenue, NW

Washington, DC 20001

www.regnery.com

Distributed to the trade by

National Book Network

Lanham, MD 20706

Manufactured in the United States of America

10 9 8 7 6 5 4 3 2 1

Books are available in quantity for promotional or premium use. Write to Director of Special Sales, Regnery Publishing, Inc., One Massachusetts Avenue NW, Washington, DC 20001, for information on discounts and terms or call (202) 216-0600.

For Max. The future is yours.
May it remain free, and full of energy.

CONTENTS

Preface   xiii

Part I: Environmentalism and Authoritarians

Chapter 1:   **Green Is the New Red: The Anti-American, Anti-Capitalist,
and Anti-Human Agenda of Today's Environmentalists**   3

Well-connected greens

Green on the outside, red to the core

People: The enemy

Your friends and neighbors

The familiar goal: Global salvation

Environmentalism as religion

Confused priorities

The betrayal of a movement

Chapter 2:   **The Authoritarian Impulse: Environmentalists Want
to Run Your Life**   25

Global warming: So sue me

Nuremberg for dissenters: "Are you now
or have you ever been . . ."

More of the same

Chapter 3:   **The Sky Is Falling**   37

Alarmism made easy

Good news is no news

Doomsay your way to fame and fortune

Global warming alarmism: Who's the greatest?

Nuke the . . . nukes

Killer . . . trees?

Part II: Global Warming: The Convenient Lies

Top Ten "Global Warming" Myths

Chapter 4:   **Global Warming 101**   65

What's causing the warming?

How much are we warming?

How destructive is this warming?

The shame

Cure worse than the possible disease

Chapter 5:   **The "Consensus" Lie**   **81**

There is "consensus" and there is truth

How dare you speak!

Al Gore's undercount

Silencing dissent

More heretics to burn

Blinded by science

Chapter 6:   **Getting Hot in Here?**   **111**

Forgetting Siberia

Cold below the belt

The United States of hotness?

Nice figures

Airbrushing the past

Hockey Stick takes center ice, denies an Ice Age

Hockey Stick in the penalty box

Are models really dumb?

The North Korean tiger

Chapter 7:   **Melting Ice Caps, Angrier Hurricanes, and Other
              Lies about the Weather**   **141**

Can't bear the truth

Chilly reception for Arctic claims

Sea no evil

Doing something about the weather

The numbers game

Hurricane Kyoto

Drought

Al Gore: Couldn't stand the weather

Part III: The False Prophets (and Real Profits) of Global Warming

Chapter 8:   **Media Mania**   169

All the alarmism that's fit to print

Full court press

Historical hysteria

It came from Hollywood

Chapter 9:   **The Big Money of Climate Alarmism**   193

Enron: Leader of the Axis of E's

"Beyond Petroleum"? How about getting beyond BP?

Chemical dependence: DuPont

No utility: Cinergy (now Duke Energy)

Big Business and Big Green

Chapter 10:   **Al Gore's Inconvenient Ruse**   209

A better title: *One Flew Over the Cuckoo's Nest*

Raging bull

Apocalypse now

Sins of omission

Sins of commission: The usual suspects

Melting ice sheets

Part IV: Making You Poorer and Less Free

Chapter 11:   **The Cost of the Alarmist Agenda**   245

Energy demand

Carbon rationing and Capitol Hill

Cost/benefit analysis: Dividing by zero

Turning out the lights

It really ain't easy being green

It's been tried

Blind optimists at home

Stick it to the poor

No benefit

Renewable boondoggles

The greater good

Poorer and no cooler

Chapter 12:   **The Kyoto Protocol**   **271**

Blame it on Rio

The Senate's unanimous rejection

Kyoto's flaws

Impotent Kyoto

Europe's embarrassment

Bad science, bad math

Bush and Kyoto

Prattling political polemics

Global governance

The Holy Grail

Conclusion   **303**

Notes   **305**

Index   **339**

✳ ✳ ✳ ✳ ✳ ✳

PREFACE

Al Gore and his friends—social, corporate, and media elites, Europeans, and UN aficionados—declare "global warming" an unprecedented global crisis. Hyped as an environmental nightmare, global warming hysteria is truly the environmentalist's dream come true. It is the perfect storm of demons and perils, and the ideal scare campaign for those who would establish "global governance" (Jacques Chirac's words in praise of the Kyoto Protocol) with strict control over corporate actions and individual behavior.

Environmentalism has served for decades as the best excuse to increase government control over your actions, in ways both large and small: It's for Mother Earth! It's for the children! It's for the whales! But standard, run-of-the mill environmental scares of the past proved to be of finite utility. Most pollution issues are relatively local—confined to individual sites or even regions. The bigger-ticket items—acid rain, the ozone hole—had been addressed and simply weren't ripe for revisiting until the next generation.

Global warming possesses no such weaknesses. Not only is planetary existence on the line, but with global warming, the greens can argue that greenhouse gas emissions in Ohio threaten people in Paris. Global problems demand global solutions, they argue, thus helping to bypass the irritating obstacles posed by sovereignty and democratic decision-making.

Upon review, however, it turns out that if global warming were as bad as they say then no policy imaginable—much less currently on the table—could "solve" it. According to the greens' own numbers, worldwide deindustrialization is absolutely critical—given available and even foreseeable energy technologies—if we are to save the planet. This explains the mantra honed during prior alarms, that "we must act now!"

Indeed, with "global warming," no matter how much we sacrificed there would still be more to do. It is the bottomless well of excuses for governmental intervention and authority.

Finally, real pollution problems can be addressed through technological improvements. Burn the fuel more efficiently and you reduce smog. Slap on a catalytic converter, and you cut down the carbon monoxide. But, as with catalytic converters, improving technology and increasing efficiency of combustion tend to *increase* carbon dioxide production. Maniacal green opposition to dams and nuclear plants—and windmills where there might be birds or a wealthy Massachusetts politician's view—ensures the only established way to significantly cut CO_2 emissions is to significantly cut energy use. What a wonderful new excuse for finally obtaining governmental—preferably supranational—control over energy. Control over energy means control over the economy and your life as you know it (as anyone who has lived through blackouts and brownouts can attest).

Governmental "solutions" to global warming would not be a matter of merely paying more to swap out some light bulbs and skipping a few trips. Al Gore has likened his crusade against CO_2 to World War II. World War II featured internment camps, food and fuel rationing, and conscription. What will Al Gore's Global Warming War entail? He won't tell us. But, after calling for a commitment on a par with a world war, or at least the Apollo moon-shot, he does intimate that it's just waiting for us on the shelf. Yet expert opinion remains clear: at the least we'll see massively higher costs and direct or indirect energy rationing. (Europe is already proving this.) At worst . . . it's scary to imagine.

This is the key to "global warming" hysteria: unless you are distracted by threats of the Apocalypse, you might question what they demand.

It is obvious that much depends on the outcome of this battle over energy and economic sovereignty, over free and open debate on science and policy—which is why the alarmists do whatever they can to avoid actual debate. They declare there is "consensus," a political concept generally alien to the scientific method. They liken skeptics to Holocaust deniers and demand "Nuremberg-style" trials of the disbelievers. They want to control our lifestyles—and they don't want you to question their cause.

This book will give you the details and the debate that they don't want you to know about.

But beware—the following facts are not acceptable in polite company. If my own experience and that of my colleagues is a guide, by uttering these (pardon the expression) inconvenient truths, you will first be accused of being a shill for evil industry. They might call you a criminal. They will suggest you gas yourself in your garage. If they ever grant the accuracy of your statements, they will warn you not to repeat them, lest you deflate the fear of global warming.

✳✳✳

The environmentalist coin has two sides, as I learned in my own introduction to this issue.

As Spring, 1991, neared I was tasked with advancing a particular environmental issue while working as a congressional fellow—essentially a glorified intern—for an up and coming U.S. Senator from the Northeast who sat on the Environment Committee. This involved legislation to regulate the lawn care and pesticide industries. Green groups were mobilized, and victims of all ages were promptly identified for a hearing, at which the horrors of these chemicals would be aired.

The night before the hearing, CBS *Evening News* helpfully presented a package on the topic. To illustrate the point—*chemicals bad*—they introduced the piece with a voice-over narrative accompanied by images of Adolf Hitler, then Saddam Hussein reviewing his missiles on parade—this being the height of Operation Desert Storm, and his WMDs not the question they are today—followed by a gentleman driving a little truck of the sort ChemLawn might trundle onto your lawn, local park, or golf course.

The subtlety and nuance left something to be desired. The aggressive public affairs campaign behind the scenes, however, did not. The entity behind this push? Another lawn care products manufacturer, one that had decided its fortune lay in marketing "green" products. They were ably assisted by their hired gun, the PR firm behind most every green scare, from Alar to global warming (and Mother Sheehan for good measure).

Several years later I left a very brief relationship with a little energy company out of Houston you may have since heard of. It turned out that I had unwittingly joined a full-scale effort to make a fortune off of advancing the "global warming" scare. Mere months after I had raised uncomfortable—and unsuccessful—questions about this internally, on August 4, 1997, company CEO Ken Lay joined British Petroleum's (then-Sir) John Browne in an Oval Office meeting with the President of the United States, Bill Clinton, and his VP Al Gore.

The agenda was to ensure the U.S. joined up to the Kyoto Protocol, an international treaty capping carbon dioxide emissions, in the name of catastrophic Manmade "global warming"—and, as it happens, to make these gentlemen an awful lot of money.

These examples do not exactly match the stories you are told. Welcome to green politics and policy.

Part I

❋ ❋ ❋ ❋ ❋ ❋

ENVIRONMENTALISM AND AUTHORITARIANS

✳ ✳ ✳ ✳ ✳ ✳ ✳

Chapter One

GREEN IS THE NEW RED

THE ANTI-AMERICAN, ANTI-CAPITALIST, AND ANTI-HUMAN AGENDA OF TODAY'S ENVIRONMENTALISTS

This is not your father's environmental movement. Your hippie uncle certainly wouldn't recognize it. While it bears the same name, and now controls the same institutions as the tree-huggers of old (as well as numerous others), its true pedigree is less green than red. Most importantly for you: environmental causes always include—and often are *primarily*—campaigns to gain more government control over the economy and individual activity. They are never fights for less control or greater liberty.

When communism didn't work out, environmentalism became the anti-capitalist vehicle of choice, drawing cash and adoration from business, Hollywood, media, and social elites. Environmental pressure groups have boomed into a $2 billion industry.[1] Much of their budget comes directly from the wallet of taxpayers through grants for public "education" and congressional schemes designed to subsidize the greens' lawyers.[2]

Spawned from the 1970s split of anti-modernists from the decades-old conservationist movement, "environmentalism" has matured into a nightmare for anyone who believes in private property, open markets, and limited government. Environmental pressure groups have no use for limiting governmental powers or expanding individual liberties. Instead, environmental claims are without fail invoked to advance the statist agenda.

Guess what?

❊ Environmentalism is big business, and greens conspire with industry to raise prices for you.

❊ Green extremists engage in "terrorist activities" according to the FBI.

❊ Wealthy capitalist countries have the best environmental performance: wealthier is healthier and cleaner.

3

Green Wisdom

"Giving society cheap, abundant energy...would be the equivalent of giving an idiot child a machine gun."

Paul Ehrlich, "An Ecologist's Perspective on Nuclear Power," Federation of American Scientists Public Issue Report, 1978

Environmentalist cries are now hackneyed staples of political rhetoric at the national level. Green lunacy has so run amok that respectable political figures (and former president Clinton) say that modern energy use poses a "greater threat than terrorism."

As with other political crusades that fail in the arena of representative democracy, the greens now see the courts and supranational bodies as their best hopes.

"Big Business" feels the heat not only from environmentalist groups but also from clever greens in the garb of institutional investors. Yet this outpouring of lucre from industry to greens is only partly a weak-kneed response to pressure, vainly seeking to buy approval through cash gifts wrapped in apologies for their chosen profession. Big Business actually promotes green alarmism in order to disadvantage competitors or gain "rents"—income from governmental policy favors. Industry and greens join forces to lobby for special preferences and mandates, sometimes to raise energy taxes and other times to limit all consumers, rich and poor, to more expensive product lines that would otherwise not have a viable market for years, if ever. It's a sweet deal that sure beats trying to make a buck through competition.

Well-connected greens

Environmentalist sanctimony has gone from simply smug to dangerously dogmatic, resembling other tragic "isms" of the last hundred years. Debate and dissent are intolerable: *No honest person could disagree with the catastrophists, therefore dissenters must be dishonest.* On the flip side, because the green cause is so noble, deception and outright falsehood are permitted means of operation.

Far from being a grassroots phenomenon driven by the scruffy teen tapping for contributions at your door, this elite-driven movement lards the coffers of pressure campaigns with wealth—commonly inherited, often corporate, and far too frequently looted from the taxpayer. Often at the first threat, industry falls all over itself to pay protection money for an elusive peace with pressure groups, only to guarantee itself regular dings for tribute and a noisy mob should the payoffs cease.[3]

The demands that greens place upon businesses extend into the smallest minutiae and the broadest business decisions. Meanwhile, green groups operate in a world free from accountability. To meet payroll, they need only to find new targets and new ways of declaring that the sky is falling.

Alger Hiss would blush at the support network of fellow travelers pushing this agenda from perches throughout domestic and international institutions. Most notable is the access the greens have gained to the wealth of the Rockefellers, the Fords, and the Sun Oil Company among

Green Wisdom

"If you ask me, it'd be a little short of disastrous for us to discover a source of clean, cheap, abundant energy because of what we would do with it. We ought to be looking for energy sources that are adequate for our needs, but that won't give us the excesses of concentrated energy with which we could do mischief to the earth or to each other."

Amory Lovins in *The Mother Earth—*
Plowboy Interview, 1977

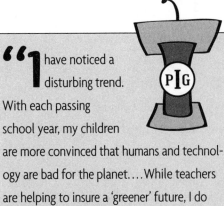

"I have noticed a disturbing trend. With each passing school year, my children are more convinced that humans and technology are bad for the planet.... While teachers are helping to insure a 'greener' future, I do not think they understand that my children may infer a condemnation of humanity."

Letter from a concerned parent to the *New York Times*, after dealing with the fallout of one too many "Earth Days," cited in *Facts, Not Fear. Discover*, October 1989

many others. Three generations removed from the entrepreneurs and businessmen who created the wealth, the charitable foundations made in their names have been perverted to wage war on the ability of today's entrepreneurs and businessmen to replicate such industrial fortunes.[4]

The green network extends to the upper reaches of supranational bodies, funded by western wealth and dedicated to redistributing—and ultimately ending—said wealth. Foremost among these is the United Nations. Consider the UN's "Global Compact,"[5] which aspires "to end capitalism," in the words of one well-placed aide speaking to a colleague of mine. The UN's population control efforts operate out of a different office.[6] Elsewhere, the UN advocates energy rationing and wealth redistribution.[7] French president Jacques Chirac praised the UN's Kyoto Protocol as "the first component of an authentic global governance."[8]

Other UN and European Union officials have made equally illuminating admissions about their aspirations for this regime, as discussed in these pages. In the words of Maurice Strong, founder of the UN Eco-Summits and undersecretary general of the UN, "Isn't the only hope for the planet that the industrialized civilizations collapse? Isn't it our responsibility to bring about?"[9]

Strong is no fringe figure, but one of the most respected and influential leaders the greens have on the international and supranational stages. Testifying beside him before the U.S. Senate Committees on Foreign Relations and Environment and Public Works, I found him to be a charming, erudite oil magnate who just happens to possess extreme views sadly representing the mainstream of the environmentalist establishment.

Green on the outside, red to the core

This raises the question of the company that greens keep and their shared bond. Communists and socialists may be environmental activists, and environmental activists may be communists or socialists, though to be one does not necessarily mean being the other. These birds of a feather do however flock together as simpatico wings of the modern Global Salvationist movement.

The political parties bearing the "Green" name have earned the nickname "watermelons": green on the outside, red to the core. In the U.S., the Green Party's agenda goes well beyond fighting pollution, and includes dramatic plans for wealth redistribution. The Green Party courted perennial Communist Party vice presidential candidate and Black Panther Angela Davis. Germany's Green Party leaders such as Petra Kelly and her ilk opposed membership in NATO, and professed to be "very tolerant" of their Communist neighbors, the Soviet Union, during the height of the Cold War.[10]

While the American media's strident *anti*-anti-communism prevents it from taking seriously any comparisons to communism, the commonality between the greens and the Reds runs deep, beyond the realms of depopulation and inhibiting individual freedoms and capital formation.

Consider that communism and anti-Americanism remain vibrant and complementary political forces in those same areas of the world where environmentalists hold their greatest sway: mainly Europe. Like old-line Reds, senior environmentalists deeply believe in the destructiveness of capitalism—in this case (despite the evidence) they believe that through capitalism we are destroying our only planet (the data tell

> ## Green Wisdom
>
> "I got the impression that instead of going out to shoot birds, I should go out and shoot the kids who shoot birds."
>
> Greenpeace co-founder **Paul Watson**, quoted in *Access to Energy*, Vol. 10, No. 4, December 1982

another story, of course). Indeed some of the most virulent, home-grown America haters such as Susan Sontag and Betty Friedan made their bones spewing environmentalist dogma, presaging the movement's future split.

In plain terms, for modern environmentalism as practiced, the enemy is capitalism. More precisely, the enemy of the modern *environmentalist* is capitalism, and environmentalism is just the chosen vehicle.

The irony of this hits home when you remember communism's environmental record. Huge stretches of the former Soviet Union have been declared "ecological disaster areas." The USSR, we now know, pumped into the ground, often near major rivers, nearly half of all the nuclear waste the regime produced over thirty years. The Reds also pumped nuclear waste into the Sea of Japan. Then there was that incident at Chernobyl, a deadly disaster unimaginable in free nations. (P. J. O'Rourke explained that Communists couldn't build a toaster that wouldn't destroy the breakfast nook.)

Regarding Communist China, the U.S. Energy Information Agency writes: "A report released in 1998 by the World Health Organization (WHO) noted that of the ten most polluted cities in the world, seven can be found in China. Sulfur dioxide and soot caused by coal combustion are two major air pollutants, resulting in the formation of acid rain, which now falls on about 30 percent of China's total land area."[11]

Facts not withstanding, environmentalists generally hew toward the larger blame-America-first crowd, ignoring and even denying that the wealthiest nation in the world has superior environmental performance to the poorer, less free nations. It is

Green Wisdom

"Us *homo sapiens* are turning out to be as destructive a force as any asteroid. Earth's intricate web of ecosystems thrived for millions of years, as natural paradises, till we came along....The stark reality is that there are simply too many of us. And we consume way too much. Especially here at home....The solutions are not a secret: control population, recycle, reduce consumption...."

Matt Lauer, MSNBC
Countdown to Doomsday, June 2006

more important to them to attack wealth and espouse the superiority of primitive, er, *indigenous* lifestyles (which our environmental elites notably elect not to live). Very few environmentalists, unfortunately, have relocated to primitive Eden unspoiled by non-indigenous lifestyles. Instead, they are disproportionately found in tony ZIP codes with vistas that simply *must* be protected from the blight of windmills.

Environmentalism was not always just a flavor of anti-capitalism. Environmental advocacy was strongly bipartisan until the early 1970s, when Rachel Carson wrote *Silent Spring*[12] and energized the crowd of chemophobes and other extremists, radicalizing the movement and putting the "mental" in "environmental." Indeed, conservatives were the forerunners of conservationism, from Edmund Burke down through Russell Kirk. The conservative philosophy about natural conservation is still summed up by the line: "Every day is Earth Day if you own the land." Nothing has changed but the movement.

To committed greens, the environment is just another demonstration that capitalism doesn't work, that too many people are consuming too much of our planet's resources, and sooner or later our planet will violently react. If capitalism is the force behind "too many people" and their access to "too many resources," then capitalism is the problem.

> ## Books You're Not Supposed to Read
>
> *Hard Green: Saving the Environment from the Environmentalists*, by Peter W. Huber; New York, NY: Basic Books, 2000.
>
> *Toxic Terror: The Truth behind the Cancer Scares*, 2nd Ed., by Elizabeth Whelan; New York, NY: Prometheus Books, 1993.

People: The enemy

It is important not to glaze over the green antipathy toward *people*. In the eyes of an environmentalist, people are pollution.

The left-wing and massively pro–nanny state UK press is wonderful in illuminating for us what our elite betters think. Consider "Attenborough:

people are our planet's big problem," citing the famed naturalist Sir David Attenborough, and Professor Chris Rapley's, "Earth is too crowded for Utopia." This latter piece on the BBC by the director of the British Antarctic Survey was nastily accompanied, as is so often the case, by pictures of squalid-living brown people in case you missed the point.

When normal humans look at another human, we see a mind, a soul, and a set of hands. The greens see only a stomach. Our species' proliferation is no small aggravation to our green friends, who long have predicted outlandish population figures and concomitant nutritional disaster, and adamantly insist that current population is "unsustainable," a claim they have been making for decades. According to doomsayers like Paul Ehrlich, the proper or "sustainable" population of the Earth is between one and two billion; above that, famine is guaranteed. Somehow, on a "starving" planet housing well over six billion, obesity is declared an epidemic.

Security Is Overrated

Green activists insist that "world leaders must not allow concern for energy security to distract them from taking promised action on global warming."

Reuters,
June 14, 2006

Despite this abhorrence of population, there is little evidence that environmentalists disproportionately depopulate themselves (notwithstanding their proclivity to chain smoke). They generally suggest instead that *others* serve as the human sacrifices necessary to save the planet.

This reality, combined with the view of people as pollution, explains why environmental groups now assess a politician's "environmental" fitness in part on the abortion issue.[13] When the League of Conservation Voters issued its 2001 scorecard, giving members of Congress a score on environmental friendliness, it counted as "pro-environment" a vote to extend U.S. foreign aid to abortion providers.

Our alarmist friends are undeterred by being proved more wrong every year. Meanwhile, not capitalism and wealth, but bureaucracy, governmental corruption, and failures to allow economic liberties impede worldwide nutrition.

The issue of green antipathy for the human race has warranted deeper treatment in many volumes. The greens' philosophy can be described by the shorthanded IPAT, or Impact (bad) = Population x Affluence x Technology (often interpreted as per capita energy use). Naturally, Ehrlich had a hand in introducing this formula, (which tells us all we need to know about its predictive value). IPAT is so reliable that, as the Cato Institute's Jerry Taylor has pointed out, it demands that Americans should be migrating en masse to Botswana, Albania, Namibia, Gabon, Laos, Armenia, Moldova, and other garden spots where IPAT scores outrank those generated by our own miserable existence here at home.[14]

Your friends and neighbors

It seems fair to say that most middle-class Americans who consider themselves "environmentalists," particularly those who are not *professional* environmentalists, sincerely believe that human development and prosperity severely harm the environment in general, and climate in particular. Busy people relying upon superficial if breathless media treatments of the issues succumb to this view although their education and experience permit them, upon the slightest scrutiny, to understand that wealthier is indeed healthier—and cleaner.

Green Wisdom

"We've already had too much economic growth in the U.S. Economic growth in rich countries like ours is the disease, not the cure."

Green guru **Paul Ehrlich**, author of the spectacularly disproven *Population Bomb and Population Explosion*, and therefore esteemed environmentalist and academic

Environmental policies come with a cost, often to the society as a whole, decreasing wealth, and so harming health. The average environmentalist, however, ignores this danger by assuming the policies' costs will fall on someone else's shoulders (multinational corporations, wealthier people, or if the environmentalist is a European: *Americans*). Environmental sympathies seem to offer cheap virtue.

American green activists confront a public that nonetheless remains more skeptical of government and state interventionism than the subjects of the European Union, who readily turn to the state once they are convinced a problem exists. This European faith in interventionism also yields a lower threshold of skepticism to alarmism. When government is already as big and intrusive in one's daily life as in Europe, shrugging one's way deeper into the morass seems small beer.

This is not to say that Americans are broadly disposed against calling on government to fix something believed to be a problem. (And we certainly harbor sub-populations prone to Euro-think.) Consider the response, or rather lack thereof, of some communities as Hurricane Katrina approached, and the apparent expectation that Washington would take care of it (and then the *post*-Katrina response of seeking a government-funded rebuild in the same storm-prone, sub-sea level location). If you expect the government to keep you safe from hurricanes while you live on the coast below sea level, you are probably willing to give the government whatever powers it claims to need to control the weather.

Still, Americans generally maintain a vibrant tradition of private solutions, including to the

"Lest you doubt the left's pieties are now a religion, try this experiment: go up to an environmental activist and say 'Hey, how about that ozone hole closing up?' or 'Wow! The global warming peaked in 1998 and it's been getting cooler for almost a decade. Isn't that great?' and then look at the faces. As with all millenarian doomsday cults, good news is a bummer."

Columnist
Mark Steyn

Nothing to Fear but Fear Itself

"**H**as it ever occurred to you how astonishing the culture of Western society really is? Industrialized nations provide their citizens with unprecedented safety, health, and comfort. Average life spans increased 50 percent in the last century. Yet modern people live in abject fear. They are afraid of strangers, disease, of crime, of the environment. They are afraid of the homes they live in, the food they eat, the technology that surrounds them. They are in a particular panic over things they can't even see—germs, chemicals, additives, pollutants. They are timid, nervous, fretful, and depressed. And even more amazingly, they are convinced that the environment of the entire planet is being destroyed around them. Remarkable! Like the belief in witchcraft, it's an extraordinary delusion—a global fantasy worthy of the Middle Ages. Everything is going to hell, and we must all live in fear."

> An increasingly rare academic who doesn't ride the green gravy train but rejects it, as fictionalized by **Michael Crichton** in *State of Fear*, New York: HarperCollins, 2004, 455.

issue of conservation.[15] As a result, American greens have a tougher row to hoe in that they must both convince the general public that the alleged problem is real and that their prescribed policies limiting individual freedoms and taking away their money are the answer.

The same faith in the government to solve any perceived environmental problem prompts European greens to also be slightly more focused on *proclaiming* disaster—which their populace will also more readily blame on capitalism. With capitalism as the cause of the alleged malady, that malady is easier for Europeans to accept as real.

This does not mean we ought to pity the plight of American greens, with their enormous fundraising advantage over their ideological opponents, sympathetic and enabling media, and the willingness to make most any claim and viciously attack heretics. Still, the relatively poor performance of American greens in enacting the agenda prompts them to appeal to the authority of Europe as proof that America is somehow misbehaving by not aping their decisions.

With scowls from our international betters guaranteed, greens find it advantageous to move most major environmental issues to the international arena. Though American resistance to a statist agenda is thereby diluted, it is not liquidated, so long as U.S. leaders remember America's long-designated role: play the adult; be the bad cop, say "*no*" to things that others feel it is their part to demand.[16]

Yet greens are persuasive in their passion. Outside of sympathetic media to carry their message, the environmentalists' greatest strength is that their adherents do really believe in what they preach—at least in their cause if not the claims. That is not to say that the greens' motives are pure or even pro-human, but a convicted missionary is far more likely to convert others, or at least persuade them of the justness of his mission.

"**S**pecial interest extremists continue to conduct acts of politically motivated violence to force segments of society, including the general public, to change attitudes about issues considered important to their causes. These groups occupy the extreme fringes of animal rights, pro-life, environmental, anti-nuclear, and other political and social movements. Some special interest extremists—most notably within the animal rights and environmental movements—have turned increasingly toward vandalism and terrorist activity in attempts to further their causes...."

Congressional testimony of then director of the FBI **Louis B. Freeh**, May 2001

Offsetting all these advantages, the greens have one great weakness: they are wrong both in the economics and science of most every issue they now pursue. Truth does eventually come out. For example, in less than a decade their zealous, anti-"genetically modified foods" campaign appears destined to finally peter out, given that there remains no demonstrated harm from crops designed to resist threats from climate and pests. Such technological advances, under way for centuries despite green mythology of futuristic "Frankenfood," are indispensable in fighting hunger. It also seems possible, despite the massive sums at stake, that over the same span of time and once the public confronts the scope of the global warming agenda that issue, too, will be a mere footnote that the greens seek to run away from as they do "global cooling."

The familiar goal: Global salvation

The green movement's radicalization pits it opposite those who recognize that "wealthier

> ## Green Wisdom
>
> "To feed a starving child is to exacerbate the world population problem."
>
> **Lamont Cole** (as quoted by Elizabeth M. Whelan in her book *Toxic Terror*)

is healthier...*and* cleaner." They rail against wealth in the face of oppressive evidence demonstrating that short of a certain level of societal (per capita) wealth, misery is ensured, and miserable environmental effects follow. Once a certain standard of living is maintained, societies increasingly can afford to "care" about the environment in the form of pollution regulations. The richer the society (and at any point in time the stronger the economy), the more stringent and numerous the environmental regulations its citizens will tolerate or even demand.

Today's wealthiest countries regulate "parts per billion" of this substance or that and spend billions of dollars in the name of *hypothetical* risks.

Despite this correlation between wealth and expensive environmental indulgences, greens worship from afar primitive lifestyles while those mired in such lifestyles would kill to escape them. (Many doubtless do and many others die trying.) The proper response to hearing of the loss of an indigenous lifestyle is "good!" Instead, greens wince at the thought of "indigenous" populations obtaining electricity, automobility, and residential comforts that these same greens would not do without. Beneath this hypocrisy is the arrogant belief that, as the enlightened, they know what's good for these poor people. That only 5 percent of Malawi has electricity is apparently good cause for Madonna to bring a Malawian child home, but not to export the horrors of our prosperity to them.

Meanwhile the greens, as with the Left generally, bemoan not poverty, but the gap between wealthy and poor, typically refusing to acknowledge that the poor (in rich countries) are getting richer such that poverty is continually being redefined upward. It is poverty that kills, not inequality,[17] but today's poor in wealthy countries have those amenities that a century ago only the truly wealthy possessed: automobility, indoor plumbing and other modern conveniences, indoor climate control, telephones and cable television, no shortage of food and even obesity. (Presumably, widespread gout among welfare recipients would be either ignored or decried as further proof of global warming.) Even the Inuit Eskimos complain, from modern homes, of a "Right to be Cold" and that "global warming" is ruining their traditional way of life, while carping about the cost of gasoline and that their

Green Wisdom

"We must make this an insecure and uninhabitable place for capitalists and their projects. This is the best contribution we can make towards protecting the earth and struggling for a liberating society."

Ecotage (as in sabotage), an offshoot of Earth First!

airport runway has buckled.[18] What's Inuit for "*chutzpah*"?

In other words, forget the increasing wealth of the poor—it's the disparity, and the wealth of the wealthy, that they hate (despite the generosity of the wealthy in free societies). That is to say they hate wealth. Again, the source of wealth is capitalism.

Consider another aspect of environmentalism's devolution to its present, knuckle-dragging stage. With the religious tradition in Europe and much of the United States despairing, two idols were advanced to fill the void in Man's need to worship, to believe, to find authority and meaning to life: the state and the environment. The author of the *Index of Environmental Indicators*, the American Enterprise Institute's Steven Hayward, cites *New Republic* columnist James Ridgeway offering one interpretation of this gravitation, as the greens' internal schism developed in the early 1970s and before the Soviet collapse: "Ecology offered liberal-minded people what they had longed for, a safe, rational and above all peaceful way of remaking society... [and] developing a more coherent central state...."[19]

With the subsequent collapse of communist regimes, environmentalism finally emerged as a major vehicle for "remaking society" through a supreme "central state."

Former chief economist for the Organization for Economic Cooperation and Development (OECD) David Henderson called this impulse "global Salvationism." One branch of the salvationists, Henderson explained, consists of "deep-green" environmentalists, who wish to assert the rights of other living creatures, and of the earth as a whole, against what they view as the damaging and destructive activities of human beings.

> ## Green Wisdom
>
> "This is as good a way to get rid of them as any."
>
> **Charles Wursta,** chief scientist for the Environmental Defense Fund, in response to the likely millions to die if DDT were banned (as quoted in *Toxic Terror*)

Environmentalism as religion

This is a topic on which much has been written. It has even been litigated as an Establishment Clause issue: do governmental regulations instituted in the name of environmental causes impermissibly intertwine the state with religion—in this case the faith of "deep ecology"?[20] The issue itself has never been judged on the merits but we can hope that, when it is, the court will give the issue respectful consideration.

This book does not attempt to recreate the well-articulated arguments on the matter. A riveting treatment of environmentalism as religion is offered by popular novelist and now enemy of the green state Michael Crichton, in his September 15, 2003, speech, "Environmentalism as Religion" to the Commonwealth Club of San Francisco.[21] For an academic treatment of the economic and regulatory aspects of this conundrum, see Robert Nelson's "*How Much Is God Worth*?"[22]

Green Wisdom

"The truth is that Mozart, Pascal, Boolean algebra, Shakespeare, parliamentary government, baroque churches, Newton, the emancipation of women, Kant, Marx, Balanchine ballet et al., don't redeem what this particular civilization has wrought upon the world. The white race is the cancer of human history. It is the white race and it alone—its ideologies and inventions—which eradicates autonomous civilizations wherever it spreads, which has upset the ecological balance of the planet, which now threatens the very existence of life itself."

Susan Sontag, *Partisan Review*, Winter 1967, 57

Confused priorities

Capitalism is the enemy but so, too, is logic, it seems. Environmentalism is riddled with so many contradictions and paradoxes that its adherents simply cannot maintain the green religion in good faith. Green hypocrisy runs far deeper than jet-setters deriding American automobility or celebrities who day-trip into Third World poverty—which motivates our Hollywood friends like little else this side of a red carpet. It extends beyond the Kennedys and Heinz-Kerrys of the world believing that wind-mills must be placed everywhere *else* than offshore their Cape Cod manses as that would despoil their view.

For example, for decades the green activists and their useful idiots in our state and federal legislatures have tirelessly worked to ensure passage of endless prohi-bitions and mandates with the inevitable and typically intended consequence of decreasing the availability or usability of "fossil fuels" such as coal, oil, and natural gas, raising the price of gasoline at the pump and ensuring rolling black-outs during periods of increased demand, electricity's equivalent of gas lines.

A Book You're Not Supposed to Read:

Eco-Imperialism: Green Power, Black Death, by Paul Driessen; Bellevue, WA: Merril Press, 2003.

Greens cry: *we need to eliminate our dependence on foreign oil… but no drilling here*, be it in Alaskan tundra or dozens of miles offshore (we have ceded the latter to Cuba and China, in our wisdom). A green mantra in response to shortages (that they engineer), absurd upon even a moment's reflection, is some variant of "our greatest proven reserve is conservation."

Though they try hard not to say it directly, it is undeniable that the demand to reduce our use of "foreign oil" means "*any* oil": *Don't use imports, but you can't drill here.* Similarly, insistence that we reduce our use of "(fill-in-the-blank) energy" means "*energy.*" There's no need

The greens and their anti-globalization comrades have never actually opposed globalization but have been rabid promoters; for years their ideological brethren vowed to export their philosophy to every corner of the world. Having lost that fight, the "anti-globalists" simply oppose globalization of *capitalism.*

to fill in the blank. On the first Earth Day in 1970 the U.S. depended about one-quarter on foreign oil. With increasing green-demanded restrictions on domestic production, ensuring that increased domestic demand is met with other-than-domestic supply, this dependence is now pushing 60 percent.

The green groups that *brought us* "dependence on foreign oil" by locking up our reserves now have actually seduced conservative hawks into canoodling with them to promote an updated version of Jimmy Carter's energy policy, ostensibly to impoverish the Middle East and thereby somehow cure terrorism.

It should now be apparent that economic illiteracy and ignorance of energy markets are threshold requirements for environmental activism.

As an example, when gas prices hit three dollars a gallon and oil companies made record profits in 2005 and 2006, why, *three dollars is an outrage!* The greens' responses were highly illuminating. Their mass-emailed missives were a dog's breakfast of xenophobia, disingenuous bemoaning of the price, and the usual global warming alarmism (their "solution" to which is the same as their "solution" to global cooling and most everything else: make energy even more scarce). Given that the environmentalist agenda is motivated by ensuring higher energy prices and reducing energy use, this smacks of cheap capitalization on an emotional issue to advance an agenda. Because it is.

To test this, consider that Europe's gas prices have run to over six dollars a gallon, but the overseas affiliates of our domestic greens (our anti-corporate crusaders being typically *de facto* if not *de jure* multinational corporations or franchises) have not raised a cry. Europe and the U.S. pay the same price for a barrel of oil. The difference in the price we pay at the pump is almost exclusively taxes (each nation has its own biases and degree of ethanol or "moonshine" boondoggles distorting the price, but the taxman's take is generally equivalent to the per-gallon price difference). With the revenue from the higher cost going to the state, and not a corporation, six-plus dollars a gallon is just fine, thank you.[23] Who can take issue with a "sin tax" virtuously contributing to the public purse?

In fact, those readers who follow the debate closely know that even the current price in Europe is far too low for the greens' tastes. Already the European Parliament is making noise about a "Kyoto tax," while spending millions on a PR campaign to convince their subjects to voluntarily eschew what automotive freedom they have. The little guy somehow got forgotten here.

Consider the policy arguments surrounding "the greatest threat facing mankind, worse than terrorism."[24] That is, for the uninitiated, "climate change." *Global warming due to greenhouse gas emissions will kill millions! But we'll never allow greenhouse-gas-free[25] nuclear power!* There is certainly a moral tension in *Millions will die...but your nuclear power, it...frightens...me.*

Dams and even tree farms could help alleviate the alleged climate crisis, but they too are apparently worse than even the CO_2-induced apocalypse. Green groups used to advocate hydropower, windmills, and nuclear power, until they started to actually appear. Dams are mean to fish. Windmills, those avian Cuisinarts, were the promised "new" (chuckle) technology if we would only swear off abundant coal. All the wind farms in the world wouldn't replace fossil-fuel plants, because wind

> ## Green Wisdom
>
> "The only real good technology is no technology at all. Technology is taxation without representation, imposed by our elitist species (man) upon the rest of the natural world."
>
> **John Shuttleworth,**
> Friends of the Earth
> manual writer, quoted
> in *Toxic Terror*

False Prophecies

"Certain signs, some of them visible to the layman as well as the scientist, indicate that we have been watching an ice age approach for some time without realizing what we are seeing....Scientists predict that it will cause great snows which the world has not seen since the last ice age thousands of years ago."

Betty Friedan,
"The Coming Ice Age,"
Harper's, September
1958

power's intermittent nature requires backup generation—typically "fossil fuels"—spinning idly at low-efficiency, below-peak levels.

In truth, not only have there been numerous "greatest threats," but the rhetoric surrounding each is increasingly revealed to include the unspoken caveat: "...*except for the others.*"

The betrayal of a movement

Naturalism wasn't supposed to mean statism. Naturalism's modern successor, today's environmentalism bordering on ecotheism, asserts that the only way to preserve nature is through state control of resources and liberties. By this thinking, the falling Iron Curtain should have revealed a green Eden, but we know that to be far from the case.

Today's "environmentalist" comes in many varieties, from ill-shaven undergrads to wealthy elites, with myriad motivators driving this activism. The modern environmentalist's motivation is generally not a love of biological diversity or horticulture, nor a desire to expand animal habitats and so on—though these advocates certainly do exist, in obscurity.

Instead, today's environmentalist is generally "anti-" something, and that *something* is typically related to growth: economic growth, population growth, physical development, or simply the individual property rights necessary for growth. Cute panda logos and other kitsch aside, outside of the rare breeds arguing in favor of one cause or another, today's environmentalist is removed in scope and degree from his naturalist ancestors.

Again according to AEI's Hayward, there are "the distinct echoes of Rousseau and his successors . . . in popular environmental thought—the view, in a nutshell, that man is estranged from a benevolent state of nature, that human society and institutions corrupt man's harmony with nature, and can be changed through a supreme act of will."[26] The Rousseauian ideal does persist in the modern environmental activist, who is equally likely to dreamily imagine that truly wild places exist, that is, places "unspoiled" by human presence.

Although environmentalists have split off from their twentieth-century conservationist predecessors, they continue to lure conservationist groups to aid them in locking up lands as "public" by beguiling them with the prospect of more free places to hunt and fish. The greens don't let on that they would make such lands "single use," which turns out to mean lying in bed knowing that these places exist, and may be visited (typically on foot). The greens have steadily lobbied to curtail more and more activities on "public" lands, beginning with anything motorized but ultimately extending to other behaviors the greens find odious, such as application of hooks or small-gauge shot to animals.

In contrasting old-school naturalists and conservationists to today's environmentalist, the twenty-first-century green begins to look not only anti-American or anti-capitalist, but nearly anti-human. An October 2006 article in *New Scientist* dreamed about a world in which all humans disappeared tomorrow.[27] The author quoted "conservation biologist" John

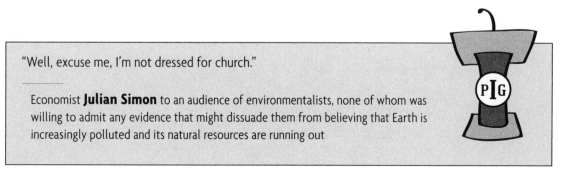

"Well, excuse me, I'm not dressed for church."

Economist **Julian Simon** to an audience of environmentalists, none of whom was willing to admit any evidence that might dissuade them from believing that Earth is increasingly polluted and its natural resources are running out

Orrock as saying: "The sad truth is, once the humans get out of the picture, the outlook starts to get a lot better."

This distaste for man did not so much develop as a *result of*, but manifested itself *in*, the advent of Professor James Lovelock's Gaia theory. That theory of the Earth as a self-regulated living being, though ostensibly scientific, is as much ideological or theological in that it places the Earth not as a creation of God but a goddess in her own right.[28]

In this world, people are pollution, bringing to reality the party mantra in Orwell's *Animal Farm* of "four legs good, two legs bad." The Reverend Thomas Malthus has overtaken Rousseau as the greens' dashboard saint, he who is annually proven so spectacularly wrong in his predictions of horror borne of scarcity in a world of finite resources and growing population. Driving the modern environmentalist are not dreams of wide open spaces but nightmares of advancing wealth, population, and technology. More importantly, enticing the modern environmentalist is the promise of central control over businesses and individuals. And who better to be the central planner than the enlightened greens themselves?

Chapter Two

THE AUTHORITARIAN IMPULSE

ENVIRONMENTALISTS WANT TO RUN YOUR LIFE

Whether you call it interventionist, socialist, or worse, there is little doubt that environmentalists throughout modern history have instilled fear over one looming "crisis" or another with the aim of increasing government control over things big and small. They see state control as a good thing in itself and pursue it aggressively and by any means necessary, because individual liberty is inherently dangerous in their eyes.

Milton Friedman noted in his 1994 introduction to Hayek's *The Road to Serfdom*, "[t]he bulk of the intellectual community almost automatically favors any expansion of government power so long as it is advertised as a way to protect individuals from big bad corporations, relieve poverty, protect the environment, or promote 'equality.'"[1] Increasingly, each of these elements is present in most environmentalist rhetoric.

If you control how crops may be grown, what sort of crops may be grown and how farmers may deal with threats to crops such as pests and weather, you control quite a lot. Insert onerous "environmental" hurdles in trade agreements to impede commerce between rich nations and poor, and this is further true. Limit use of private property and advance other restrictions through "smart growth" policies, and the control increases further. The list goes on with how the Green Left limits how large a car you may drive, in which lanes you may drive, and how big and where your house may be.

Guess what?

- Environmentalism is usually an excuse for more government power. (Has it ever been an excuse for less?)
- Leading greens oppose affordable energy.
- Environmentalists want to prosecute those who disagree with them.

"The activists, many claiming to be associated with Friends of the Earth, circulated among the villagers before the food was distributed. One activist from Brazil was particularly shameless in his tactics. He kept telling several village women over and over that the food was 'contaminated' and 'toxic' and would harm their children."

Ron Bailey, *Reason* magazine, September 17, 2003

This is all piddling nanny-statism, however. If you want real power, you need to dictate acceptable energy supplies and consumption by claiming it is necessary in order to save the human race. Then you're talking about total control over the whole economy.

Simon Jenkins updated Milton Friedman's warning recently in the *Sunday Times* (UK):

All panics are equal. But some are more equal than others. Present-day government warns us to be very, very afraid, successively of AIDS, Saddam Hussein, BSE, terrorists, SARS, bird flu and now global warming. Rulers were once elected to free us from fear, not to increase it. Now they cry wolf every day and use it to demand more power and money into the bargain. Climate change is a hell of a wolf. Last week the BBC's resources were marshaled to produce a royal variety performance of usual suspects: retreating Patagonian glaciers, collapsing Arctic ice shelves, starving Africans, burning rainforests and storm-lashed New Orleans. It was the best of the end of the world, meant to scare us witless.[2]

Forget for the moment the hideous specter, in the eyes of most environmentalists, presented by personal property rights and open markets. The greens' latest fetish is avowing an agenda unavoidably aimed at increasing the cost of energy use, while also ultimately shifting energy policy from the sovereign to the multinational level (where they have the greatest influence and, not surprisingly, least accountability). Until such time as the international global warming regime can be empowered, clever greens in the UK have called for a "third way" of taking the issue out of the democratic process and handed over to an "authoritative independent body" to institute the desired mandates.[3] It seems that public resistance to the lifestyle police is getting in the way.

Posturing may indeed be a significant act in the green repertoire, but no one will accuse the movement-greens of being in it merely for show, of not being committed (even though "commitment" might be in order for many of them).

Whatever Happened to "Keep Your Laws Off My Body"?

Four decades ago, scientists were so determined to prevent famines that they analyzed the feasibility of putting "fertility control agents" in public drinking water. The physicist William Shockley suggested using sterilization to impose a national limit on the number of births.

Planned Parenthood's policy of relying on voluntary birth control was called a "tragic ideal" by the ecologist Garrett Hardin. Writing in the journal *Science*, Hardin argued that "freedom to breed will bring ruin to all." He and others urged America to adopt a "lifeboat ethic" by denying food aid, even during crises, to countries with rapidly growing populations.

Those intellectuals didn't persuade Americans to adopt their policies, but they had more impact overseas. Under prodding from Westerners like Robert McNamara, the head of the World Bank, countries adopted "fertility targets" to achieve "optimal" population size. When an Indian government official proposed mandatory sterilization for men with three or more children, Paul Ehrlich criticized the United States for not rushing to help.

John Tierney, "The Kids are All Right," *New York Times*, October 14, 2006

If the claims of imminent doom were sincere, however, and the greens believed they could back them up (that is, were the science "settled"), it certainly seems they would demand something other than Kyoto or its watered-down progeny circulating through the U.S. Congress. That is, if global warming were as bad as they say it is, the proposed "solutions"—in the form of energy limits, manufacturing controls, and other freedom-restricting and economy-crippling measures—would need to be *much* more drastic than those proposed. That would force a debate, of course, on the reality of the threat necessitating such draconian intervention; but the true believers should have no problem with that. While they reveal fairly modest "first steps" the greens have much bigger—okay, smaller and fewer—things in mind for you.

The greens have learned to merely seek the dose of the poison that seems politically acceptable before moving on to the next step. This is why the "global warming" campaign is so insidious: once one buys into the threat and need to act, and accepts the costly but impact-free demands, one can hardly then object to actually "doing something" in the form of the next twist of the noose. See the discussion of Rio and Kyoto in Part IV, demonstrating that sometimes the greens miscalculate.

With control come conundrums. Power over your daily life is purportedly a means to the end of preventing climate change. Thirty years ago, they needed to limit economic activity in order to prevent global cooling. Today it is global warming that justifies increased governmental nannying and restrictions. But consider the green predicament were

"Every citizen is given a free annual quota of carbon dioxide. He or she spends it by buying gas and electricity, petrol, and train and plane tickets. If they run out, they must buy the rest from someone who has used less than his or her quota."

Proposal by British columnist **George Monbiot**, October 31, 2006 presaging what has since been leaked as a government proposal

they offered the power to set the Earth's thermostat:[4] where would they set it? The answer would prove to be "nowhere," as the ultimate sin is to tamper with nature.

The control they seek is over you, not *Her*. What, if any, interference would the environmentalists brook to stop global warming (or cooling) if it were determined that global warming was entirely due to natural, non-human, causes? What if the predictions of rising temperature were found to be accurate, but the cause was found to be something other than that dastardly *Man*? What should we do then to prevent global warming?

Again, nothing. We know this because according to our greens the natural state—the mythical natural stability of climate—is the ideal: wherever it is very cold, that is good; wherever it is very hot that, too, is very good. It's a bit of the Goldilocks syndrome: in the '70s it was our fault and too cold, in the '90s it was our fault but too hot; only the world untouched by Man is just right. There is no perfect ambience except that which is natural, and things can only be "natural" by enabling the bossy green establishment to tell you how and where to bug off. (That they

On August 1, 2006, the BBC reported that "efforts to get households to reduce energy use are being hampered by" family breakdown, adding unnecessary housing units ("appliance packed households of single men") consuming energy.

The greens will only go so far to promote their agenda, however, so instead of encouraging marriage the proposed responses were collective housing, requiring eco-friendly homes, and an occupancy tax.

The prior month, news broke of a plan under consideration by the UK government to issue each citizen a "carbon swipe card" to ration energy use.

presently implicate Man in major weather events has turned the term "natural disaster" into an oxymoron.)

Global warming: So sue me

As with abortion on demand, homosexual marriage, and some other agenda items of the far Left, the fight to micromanage you in the name of global warming may find its best friends in unelected judges and unaccountable international tribunals.

Many global greens touted the International Criminal Court as an "environmental" treaty.[5] Similarly, they continue clinging to white elephants such as the failed "Convention on the Protection of the Environment through Criminal Law."[6] Establishing an international environmental plaintiffs' bar is a priority for these same greens.[7] All of these efforts reveal how some environmentalists see global warming as "the next tobacco." That is, they hope to use lawsuits to force industry to cede control and profits.[8]

A November 2005 Reuters story headlined, "UN examines prospect for climate-change litigation," laid the agenda out plainly:

Companies which contribute to climate change will increasingly face legal action, law firm Freshfields said on Wednesday, launching UN-sponsored research which highlights

"**W**arming (and warming alone), through its primary antidote of withdrawing carbon from production and consumption, is capable of realizing the environmentalist's dream of an egalitarian society based on rejection of economic growth in favor of a smaller population eating lower on the food chain, consuming a lot less, and sharing a much lower level of resources much more equally."

Then UC Berkeley professor **Aaron Wildavsky** on what he termed "the mother of all environmental scares"

investors' environmental responsibilities. "Twenty or thirty years ago you were looking at the beginning of tobacco litigation," Freshfields lawyer Paul Watchman said. "There's going to be a whole host of (climate-change) actions ... we might look to do that kind of thing."

"Western environmentalists ... are interested in the environment only insofar as they can exploit it as an issue to attack liberal [meaning free] societies."

French philosopher **Jean-Francois Revel**, *Anti-Americanism* (2003)

At present of course there is no chance of successfully suing a company for purportedly causing or contributing to "global warming." As also discussed in Chapter 9 this is not for a lack of CEOs willing to go along with the radical green agenda in the hope of partaking in the windfall that Kyoto-style policies would confer, in the short term, on a few.

The problem for those who would sue is that there is no law anywhere in the world against producing CO_2 or contributing to "global climate change." The "binding, enforceable" Kyoto Protocol is no such thing[9] and, even if it were, it does not provide for forcing payments out of individual

Green Wisdom

"The right to have children should be a marketable commodity, bought and traded by individuals but absolutely limited by the state."

Kenneth Boulding, originator of the "Spaceship Earth" concept (quoted by William Tucker in *Progress and Privilege*, 1982)

No Plumbing for You!: The Green Toilet Nazis

"Introduction of the flush toilet deplored at Earth Summit"

Headline, **CNSNews.com**, reporting comments made by green leader, August 30, 2002

defendants. Instead, plaintiffs must prove that someone caused global warming which in turn caused them harm. But in lawyer-speak, "climate change" is not a viable tort, because *causation* remains impossible to establish, as does harm to any particular individual. In other words, we have no way of proving that anybody did anything that harmed anybody else. Apportioning liability and damages further muddies the waters. Who can demonstrate he was harmed by the climate change caused by a particular party—and how could this be sorted out from the wealth of positive and negative effects from the same climate change? The work of James Hansen (much more on him later), revered as the "father of global warming," suggests that, if we really

"As World Warms, Legal Battles Brew."

"Heatwaves, droughts and rising seas are likely to spur a spate of hard-to-prove lawsuits in the 21st century as victims seek to blame governments and companies for global warming, experts say. Pacific islanders might sue to try to prevent their low-lying atolls from vanishing under the waves, African farmers could seek redress for crop failures or owners of ski resorts in the Alps might seek compensation for a lack of snow. 'If the evidence (that humans are warming the globe) hardens up, as it may well do, then it has all the ingredients of the tobacco case,' said Myles Allen, of the physics department of the University of Oxford in Britain."

Reuters, July 26, 2006

are causing drastic global warming, we might just be forestalling a future ice age. Such a benefit if true is incalculable.

Also, even were courts to accept Man-made climate change, consider the trade-off between societal betterment from wealth creation and abundant, affordable energy and the alleged marginal contribution to natural climate cycles. While this truncates a topic worthy of several law review articles,[10] it is fair to say that significant problems exist for the individual plaintiff seeking damages, which helps explain why greens generally have eschewed taking the issue on in a real court,[11] despite claims of "slam dunk" science and catastrophic damage to date.

> ## No Soup for You!: The Green Food Nazis
>
> In the impoverished Mexican village of Valle Verde, green activists assailed hungry villagers being offered food donated by the Committee for a Constructive Tomorrow (CFACT), threatening that danger lurked in the items from local store shelves such as corn meal, cooking oil, beans, and Kellogg's cornflakes, some of which contained ingredients made from genetically enhanced crops like corn and canola.

Further, despite a lawsuit by crusading state attorneys general, there is no injunctive relief a court could (credibly) order. That is, there is nothing anyone could be forced to do that would prevent or reverse even one degree of climate change, whatever the cause.

Regardless, the past few years have proved that absence of the necessary elements of a tort (much less a crime) will not stop ambitious officials like New York's Eliot Spitzer from using the issue to impose great costs and extract politically charged settlements from targeted industries. Of course, for some British Leftists, the case that burning fuel causes environmental catastrophe is so "settled" that litigation is not a consideration; instead, instant justice is in order. Consider BBC favorite and Guardian columnist George Monbiot: "[E]very time someone dies as a result of floods in Bangladesh, an airline executive should be dragged out of his office and drowned."

Nuremberg for dissenters: "Are you now or have you ever been..."

If emitting carbon dioxide and methane is a "crime against humanity" (if not an actual violation of the criminal code...yet), then what are we to do about those who aid and abet this crime? Already, businesses have been subject to intimidation campaigns for daring to support any group advocating policies that either support business or resist government expansion. This is certainly the case with climate change, with even the UK's Royal Society seeking names and promises of excommunication of groups daring to dispute proclamations by the Royal Society.[12] The RS's motto, by the way, is the putative expression of skepticism "*Nullius in verba*," loosely translated as, "Don't take anyone's word for it." Well, there's skepticism, and then there's disagreement with the Royal Society.

But it's far more serious than that. Manmade global warming is a crime against humanity, the alarmists charge. It is therefore only fitting to have "Nuremberg-style" trials for those dastardly dissenters who resisted draconian government restrictions on people's energy use.

In September of 2006, the website for *Grist* magazine, which is mainstream "green" enough to be granted interviews by Al Gore and PBS's Bill Moyers, linked to an article in the UK's *Guardian* about the "denial industry," which prominently included a group I am affiliated with, the Competitive Enterprise Institute. Accompanying the link, *Grist* staff writer David Roberts wrote, "we should have war crimes trials for these bastards— some sort of climate Nuremberg." (Now, on top of my food-taster and car-starter I guess I need to dig deeper and spring for counsel

---

"Woman Is Cleared of Failing to Recycle"

"Britain's first prosecution for failing to recycle household waste has failed after a woman was cleared of putting the items in the wrong bin. Exeter City Council pledged to continue chasing recycling offenders through the courts, despite yesterday's landmark ruling."

The Independent, July 11, 2006.

---

specializing in war-crimes defense. Maybe after the Balkan mess is tidied up.)

Grist was not alone in calling for these new sedition laws. Mark Lynas—whose book *Rising Tide* begins by pointing out how rainy England seems to be these days and concludes by blaming George Bush—wrote on his website in May of 2006:

> I wonder how future juries might view the actions of the Competitive Enterprise Institute, who, in full knowledge of the realities of climate change, continue to preach their gospel of denial in the service of Big Oil dollars. I wonder what sentences judges might hand down at future international criminal tribunals on those who will be partially but directly responsible for millions of deaths from starvation, famine and disease in decades ahead. I put this in a similar moral category to Holocaust denial—except that this time the Holocaust is yet to come, and we still have time to avoid it. Those who try to ensure we don't will one day have to answer for their crimes.[13]

Scott Pelley of *60 Minutes* got the memo, too. When someone from the *Columbia Review of Journalism* asked him why he interviewed alarmists, but not skeptics, on the issue of global warming, Pelley responded, "If I do an interview with Elie Wiesel, am I required as a journalist to find a Holocaust denier?"

Working out the logic of this twisted analogy would probably leave the reader dumber than he started.

Green Wisdom

"No matter if the science is all phony, there are still collateral environmental benefits" to global warming policies…. "Climate change [provides] the greatest chance to bring about justice and equality in the world."

Canada's environment minister, **Christine Stewart,** comments at a meeting with the editorial board of the *Calgary Herald, Financial Post* (Canada), December 26, 1998

Chapter Three

❋ ❋ ❋ ❋ ❋ ❋

THE SKY IS FALLING
THE CONSTANT (AND CONSTANTLY CHANGING)
ALARMISM OF THE ENVIRONMENTAL MOVEMENT

Even Al Gore's former Senate aide and handpicked administrator of the U.S. Environmental Protection Agency, Carol Browner, regularly admitted the steady improvement of the environment as measured by many key indicators. Her final Annual Performance Report acknowledged long-term trends:

> Between 1970 and 1999, total emissions of the six principal air pollutants decreased 31 percent....These improvements occurred simultaneously with significant increases in the nation's population, economic growth, and travel and are a result of effective implementation of clean air laws and regulations, as well as enhancements in the efficiency of industrial technologies.[1]

People were traveling more (vehicle-miles traveled were up by 140 percent), the U.S. Gross Domestic Product was up 147 percent, and U.S. population climbed 33 percent. Still, air pollution fell. For eight years, Clinton-Gore took credit for these gains—which had been under way for decades—while they simultaneously claimed tens of thousands of annual deaths from air pollution whenever such claims could justify greater regulatory controls.[2]

Guess what?

❋ Greens predicted global starvation, drought...and global cooling...by the 1980s.

❋ Environmental indicators continue to improve.

❋ Environmentalists oppose clean energy sources such as nuclear and hydro-electric power.

Somehow, though, environmentalist groups and the media woke up on the rainy day of President George W. Bush's inauguration and found that there was no good news. Things were getting worse, every day, we were told, every day. This gloomy drumbeat for years doubtless laid the foundation for what Gallup concluded in its 2005 annual "Earth Day" poll.

That March, 63 percent of Americans polled said the environment was getting worse, even though the positive, long-running positive trends continued during the Bush administration. While the environment was improving, the quality of environmental *reporting* and rhetoric was rapidly declining. In fact, to read the Associated Press accounts that year, three-quarters of the country possessed the "nation's worst" air quality (more on this, below).

This environmental improvement, so upsetting to reporters and power-hungry bureaucrats, persists worldwide, as AEI's Hayward establishes in detail in *The Index of Leading Environmental Indicators: The Nature and Sources of Ecological Progress in the U.S. and the World*.[3] Others have also made the case persuasively, including Bjorn Lomborg, former Greenpeace member and generally Central Casting ideal for a loopy environmentalist.

"Scientists who want to attract attention to themselves, who want to attract great funding to themselves, have to (find a) way to scare the public . . . and this you can achieve only by making things bigger and more dangerous than they really are."

Petr Chylek, professor of Physics and Atmospheric Science, Dalhousie University, Halifax, Nova Scotia, commenting on reports by other researchers that Greenland's glaciers are melting. *Halifax Chronicle-Herald*, August 22, 2001

Lomborg proved a misfit in the green movement, willing to change his beliefs if his information changed: that is, his beliefs were driven by data, not the other way around. Teaching statistics at Denmark's University of Århus, Lomborg encountered the work of Julian Simon (the optimistic anti-Paul Ehrlich) and sicced his best pupils on Simon's rosy arguments about Man's ingenuity and resourcefulness and their impact on the environment and human health, generally. So surprised by what they found, to his eternal credit, Lomborg risked his entire social, political, and possibly professional life by valuing the evidence over the party line.

Lomborg described "the Litany" with which we are inundated, that "[o]ur resources are running out, the population is ever growing, leaving less and less to eat. The air and water are becoming ever more polluted....The world's ecosystem is breaking down."[4] He discovered to his and his students' surprise that these cries "[do] not seem to be backed up by the evidence."

Not only are air pollutants dramatically down over decades and continuing their decline, but the statistics of the loss of forest land from clearcutting are exaggerated. Lomborg dared say that real, deadly, solvable threats receive short shrift because of money squandered on green handwringing exercises.

Worst of all, Lomborg concluded that global warming would likely be rather slight. Even if warming is real to a detectable degree (which he accepts), he argues that it is nonetheless unavoidable barring massive and devastating deindustrialization. He lamented how his erstwhile allies obsess over ineffective, wasteful, and even frivolous regimes like the Kyoto Protocol that would incur massive human consequences. Sealing his fate, Lomborg dared not only deny that global warming would be catastrophic, but he asserted it would have many benefits. Thus began Lomborg's introduction to physical and verbal assaults and generally the nasty green left, unplugged.

Alarmism made easy

It seems the easiest job in the world would be "director of research" for an alarmist environmental group, given their propensity for simply making things up. (The one exception is of course Greenpeace, for whom that title implies that you're the one going through my trash every Sunday night.)[5]

The tough part for them is that they still must somehow base their stories on things that are really happening. Greens are remarkably hard-wired to turn every conceivably relevant occurrence into proof of their current alarm. This leaves them flummoxed into even calling a dramatic decline in pollutants an *increase*, or to styling a regulation of some previously unregulated pollutant as a "rollback."

A combination of the absurdities occurred when President Bush introduced his "Clear Skies Initiative." This set of regulations aimed to reduce sulfur dioxide emissions by a projected 73 percent, mercury by 69 percent, and nitrogen oxides by 67 percent. Natural Resources Defense Council and their cohorts decried this policy as an *increase* of pollution. They charged that Bush's policy would *triple* mercury emissions and *increase* sulfur by 50 percent. You see, the greens were advocating even stricter rules promising even larger emission reductions, enabling them to characterize "Clear Skies" as an "increase," if only relative to the plans they had drawn up in their fancy offices.

When President Bush proposed the first-ever regulation on mercury emissions from power plants (something Bill Clinton talked about, but never did), the

Green Wisdom

"To capture the public imagination, we have to offer up some scary scenarios, make simplified dramatic statements and little mention of any doubts one might have. Each of us has to decide the right balance between being effective, and being honest."

Climate alarmist **Stephen Schneider**, *Discover* magazine, October 1989 green groups similarly compared it unfavorably to some idea Bill Clinton's bureaucrats theorized about, but never tried to create. In these anti-Bush campaigns, of course, the greens enjoyed full cooperation of the media.

Consider smog-inducing ozone levels plummeting over the three-year span of 2003 to 2005, such that these were the three lowest-ozone years on record.[6] Conveniently for dishonest alarmists, 2005 was only the second-lowest ozone year on record since nationwide monitoring began in the 1970s, while 2004 was the lowest. The low ozone levels in 2005 were particularly good news, because 2005 was one of the hottest years on record, and, all else being equal, hotter weather tends to be associated with higher ozone levels due to the chemistry of ozone formation.

For the environmental establishment this news had the nasty odor of progress, which is particularly unacceptable under a Republican Congress *and* executive. Something had to be done.

AEI's Joel Schwartz described the reaction: "Clean Air Watch proclaimed 'Smog Problems Nearly Double in 2005.' Pennsylvania's Department of Environmental Protection warned 'Number of Ozone Action Days Up from Last Year.' And EPA's New England regional office noted that 'New England Experienced More Smog Days during Recent Summer.'

False Prophecies

"The battle to feed humanity is over. In the 1970s and 1980s hundreds of millions of people will starve to death in spite of any crash programs embarked upon now."

Paul Ehrlich, *The Population Bomb*, 1968

Writing on 2005 ozone levels in Connecticut, a *New York Times* headline warned 'A Hot Summer Meant More Smog.'"[7]

Schwartz has documented this unseemly practice as routine across the spectrum of air pollution issues (though it is by no means limited to air pollution). It is actually an annual media event: the American Lung Association's *State of the Air* report. ALA was once an iconic public health advocate which now, like others, has turned into a fundraising machine and key player in the green social engineering and fear-mongering agenda. For example, ALA promiscuously hands out "F" grades for air quality to large swaths of the country, causing the press in about half—and sometimes more—of the country to claim that theirs is the area with America's "worst" air quality. Sometimes large counties such as San Diego County are told their air is the "worst," while in fact only one small town with 1 percent of the county's relevant population is in violation of the standard.[8]

This is neither science nor public health advocacy. It is shysterism at its worst.

Cherry-picking data is not just a favorite weapon in the arsenal of alarmist activists, but it is also a beloved pastime of bureaucrats. The law requires that a regulator demonstrate that his desired proposals—with their concomitant increase in budget and authority—would provide, say, a "significant health benefit," with an "adequate margin of safety" no less.[9]

Though this threshold is a slight burden for the truly determined regulator, the EPA now brazenly and routinely selects the interpretation—sometimes the *only* such interpretation—of a set of data that can jus-

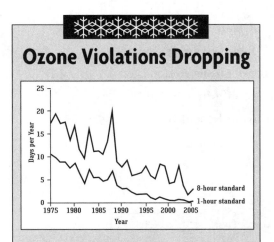

Ozone Violations Dropping

Days per Year

8-hour standard
1-hour standard

1975 1980 1985 1990 1995 2000 2005
Year

By two measures used by the federal government, U.S. cities are cutting the average number of days per year in which the ground-level ozone violated air-quality standards.

Source: EPA Air Quality System

tify its power grab. If a study can possibly be read to show a significant health benefit from regulating, the EPA will read it that way.[10] Regulators and activists routinely ignore the weaknesses in their favored studies and contrary evidence from other peer-reviewed research that doesn't support the alarmists' agenda.

In addition to cherry-picking, EPA bureaucrats and other fearmongers manipulate the evidence in a wide range of ways, for example by simply funding huge numbers of scientifically shoddy studies. Many EPA studies, for example, are *observational*, as opposed to *experimental*. Simply put, these studies try to find a correlation between some sort of malady and a particular pollutant or emission. Showing a correlation, however, is hardly a proof of causation (In a famous example, a study of England found that high concentrations of churches correlated to a high concentration of crime. If correlation equaled causation, churches were a cause

---

"In the twenty years since the Chernobyl tragedy, the world's worst nuclear accident, there have been nearly [FILL IN ALARMIST AND ARMAGEDDONIST FACTOID HERE]."

Greenpeace press release, prematurely released, May 24, 2006[11]

✴ ✴ ✴

They drew up a press release to steer

Reporters to base their stories on fear.

But imagine their surprise

When Greenpeace was apprised

They'd forgotten to put "[ALARMIST FACTOID HERE]!"

Jon Sanders, John Locke Foundation, posted on The Locker Room blog[12]

---

of crime. More likely, in this case, the two had the same cause: high population density.)

In the past, this sort of correlational study in the medical field led to claims that hormone replacement therapy and Vitamin A supplements reduce cardiovascular risks; that a low-fat diet reduces cancer and cardiovascular risks; and that calcium supplements reduce the risk of osteoporosis. All of these claims were overturned outright or drastically scaled back when real scientific method—randomized controlled trials (the more rigorous type of study used to determined whether pharmaceuticals are safe and effective)—belied the observational claims.

In other words, correlation is not necessarily causation, and statistical legerdemain is no substitute for rigorous experimental techniques. Yet these invalid observational epidemiology studies form virtually the entire justification for the continued tightening of air quality standards.

Consider asthma cases, which are going *up* despite air pollutants going way, way *down*. One could draw many inferences from this set of data, but certainly *not* that air pollution is the culprit; indeed it is more reasonable (though certainly unscientific) to conclude even that *declining* pollution causes asthma. The regulator's instinctive response, however, is the one

"No area outside California comes anywhere close to having 'some of the worst air pollution in the nation.' And yet a search through newspapers both large and small reveals that journalists and environmental activists have collectively put more than half the country into this category."

Joel Schwartz, citing dozens of newspapers racing to the bottom to claim "worst" status for their city or region, in "Air Quality: Much Worse on Paper than in Reality," *AEI Environmental Policy Outlook*, May-June 2005

we now see occurring, which is to twist the data and squander billions more to expand regulatory fiefdoms. *If air quality doesn't cause asthma*, the regulator might worry, *my agency has a less compelling case to further regulate the activities that affect air quality.*

As luck or design has it, many environmental regimes, including Kyoto and its progeny, set technically impossible standards and goals, which protects against the risk that industry might actually *meet* the standards and free themselves from the regulatory reach of the bureaucrats.

False Prophecies

"Hundreds of millions of people will soon perish in smog disasters in New York and Los Angeles...the oceans will die of DDT poisoning by 1979...the U.S. life expectancy will drop to 42 years by 1980 due to cancer epidemics."

Paul Ehrlich,
Ramparts, 1969

This is particularly true in the case of the greens' effort, argued before the Supreme Court in November 2006, to have carbon dioxide declared a "pollutant" under the Clean Air Act on the basis of "global warming." To obtain this result the court must agree that this ubiquitous, odorless, colorless, and benign gas, the overwhelming majority of which is produced by Mother Nature including by *human exhalation*, "endangers public health or welfare." This then requires establishing a "National Ambient Air Quality Standard" (NAAQS) regulating atmospheric concentrations of CO_2 at some level to purportedly cure global warming, such that the U.S.—on our own, mind you—cripples our economy by using a law never intended for rationing energy to do just that. Given the utter impossibility of ever meeting such a CO_2 NAAQS standard, this would at first move nearly all known forms of auto transport off our roads, and then, as a matter of law, demand that this regime extend economy-wide.[13]

This inescapable result is not only fine with the greens, it is at present their most aggressive project, as the sheer number of briefs filed with the Supremes pushing this agenda attests. The Sunday before the case was argued, the *Washington Post* ran three items aimed at reminding our robed wonders of the Georgetown cocktail crowd's preferences: an alarmist piece on the perils facing the ski industry, a shrill op-ed by Gore acolyte Laurie David demeaning the idea that some might actually disagree with her and Big Al, and a *Post* editorial offering the plaintiffs' legal case, as plain truth and in a vacuum. The court's opinion in this critically important case should come down at the same time as this book hits the shelves.

With bureaucrats, elites, and greens so motivated to sand the gears of our economy, we would need a vigilant and knowledgeable media or Congress to combat them. Clearly, we don't have either. Alas, our last line of defense against broken-record alarmism is common sense. The green activist mostly must worry about how to disguise the established

The Root Causes of Climate Alarmism

"I believe there are three factors now at work.

First, the discourse of catastrophe is a campaigning device being mobilised in the context of failing UK and Kyoto Protocol targets to reduce emissions of carbon dioxide.

Second, the discourse of catastrophe is a political and rhetorical device to change the frame of reference for the emerging negotiations around what happens when the Kyoto Protocol runs out after 2012.

Third, the discourse of catastrophe allows some space for the retrenchment of science budgets.

It is a short step from claiming these catastrophic risks have physical reality, saliency and are imminent, to implying that one more 'big push' of funding will allow science to quantify them objectively.

We need to take a deep breath and pause."

Mike Hulme, director, Tyndall Centre for Climate Change Research

and generally well-understood—if easily forgotten (as the 2005 Earth Day poll attests)—reality that things are getting better all the time, and convince the public that things are currently awful, or, more likely, they will be awful just around the bend. This is only possible with a media equally vested in declaring alarm.

Good news is no news

Newscasts commonly parody their own breathless announcements of a breaking *critical* threat about which you *absolutely have to know*—if you tune in at eleven, that is, or at least after this message from our sponsors. Sort of loses the urgency, no? This is not dissimilar to global warming alarm. Kyoto proponents wail about the purported Manmade climate catastrophe already with us in the here and now, though not one of them will dare offer a policy "solution" that would have a detectable influence on climate under any set of assumptions. *We must act now! And by acting I mean do nothing, except ding your incomes and lifestyles.*

The environmentalist movement, however, is committed, *über alles*, to ignoring inconsistencies and inconvenient truths. For those who can only care about the present (or the Day After Tomorrow) the greens scream about the circumstances harming or at least *with us* today: *Hottest summer! Wettest autumn! Hurricane Katrina!* For the sake of those among the public who pay attention to history, the state of their surroundings, or actual data, the alarmists alternatively insist that *real* disaster lies just over the horizon. Clearly, this is not a sustainable trend (to borrow a favorite term of the greens). Proclaiming the end of the world has a bill that simply must come due (particularly with us living much longer despite the horrors of modernity).

Wealthy societies now spend millions and sometimes billions chasing phantoms, pursuing that last molecule now even down to "parts per billion." If the EPA can detect it, they have to chase it. Marginal pollution must be removed to levels approaching natural "background" levels. This explains why the greens cry havoc the way most of us don clean underwear. But why do their otherwise intelligent cheerleaders in the media abet them? How can environmentalists get away with being the only industry publishing screeds about the looming end of the world, enclosing a subscription card for next year's issue?

False Prophecies

Climate change "will claim hundreds of thousands of lives."

Stephen Tindale, executive director, Greenpeace, quoted in *The Guardian*, March 3, 2006

I was informed of one rationale, related to a packed room gathered to listen to prominent left-of-center environmental writer Gregg Easterbrook. His personal bugaboo being the media's habitual denial of the constantly improving state of the environment, he inquired of the *New York Times* about their practice of enabling what in any other context are called lies. He was told that, were the *Times* and their media brethren to not perform this journalistic function—misleading us about the state of the environment—"the public would become complacent."[14]

Got that? You can't be trusted with the truth. The elites know what's best for you, even if you fail to see it.

As discussed in more detail in Chapter 8, the same day the *New York Times* ran its correction to its preposterous and debunked front-page story in 2000, "Unprecedented Polar Melting" (it turned out they were fed a line by a single source, an alarmist, and that in fact water at the North Pole is called "summer"), an Easterbrook op-ed appeared in the same Grey Lady. That piece began "That north pole ice has turned liquid may be the least of our problems . . . ." What the hell. No one reads corrections anyway. Later on, Easterbrook cited a study arguing that reducing methane emissions might be easy and very effective at warding off global warming. He noted the one downside of this promising development: "many environmentalists worry that any lessening of fears regarding fossil fuels will hurt efforts to reduce global warming."

Reducing greenhouse gases is good, unless it comes at the cost of "lessening fears." And "abating global warming" is good, unless it means you get to keep your minivan.

Similar to the distaste for the "wrong kind" of data is the widespread green tenet that the truth is an article of faith: *It isn't what you can demonstrate, it's what I believe, and someday I will be able to prove it. In the meantime, I will torture data until it confesses and converts to my worldview.* This faith has reduced elite environmentalism to little more than a white-collar version of the loon strolling around Lafayette Square, outside the White House, in a sandwich board demanding that you *repent now, the end is near.* (As we have seen, sometimes that green loon also is elected to prowl the halls *inside* the White House, before gaining weight and going on to make independent movies about things melting.)

Doomsay your way to fame and fortune

In May 2006 the European Parliament honored world-class alarmist and failed prognosticator nonpareil Lester Brown, founder and longtime president of the WorldWatch Institute.[15] Brown's continued prominence is most amazing given that it may be difficult to find a prediction he has made that proved correct. On that occasion when he stumbles into being right it generally has nothing whatsoever to do with his gloomy Malthusian thesis but government stupidity. And, as with most elites, it is not government with which he has a problem but the governed.

Yet, like *Population Bomb* and *Population Explosion* author Paul Ehrlich, Brown slithered upward in the appropriate professional, social, and academic circles by satisfying the elites' insatiable desire for gloom. Brown dined out for over a decade on promises that next year was the year there would finally be a net drop in total world food production and star-vation would begin to creep across the planet.[16] (Meanwhile, we see an epidemic of obesity, and the greatest hindrances on our food supply are institutions—such as the European Union—which thwart genetic modifi-cation of foods to make them hardier to better resist pests and weather.)

In late 2006, Brown released a report through his latest venture, Earth Policy Institute, suggesting that more than 250,000 people who fled the Louisiana and Mississippi coasts during Hurricane Katrina will not return to the region because changing environmental conditions pose too great a risk for rehabilitation. Quoth the dark sage Brown, "The interest-ing question becomes, 'When do hurricane evacuees become climate refugees?'" Turns out, through green lenses the answer is fairly simple—when the evacuees decide not to go back.

My sometime colleague and a longtime expert on environmental pol-icy, R. J. Smith, put it this way: "Brown discovered that there are things like violent storms and floods and heat waves and cold spells—all of which kill people—and some of which force people to move. And those events almost always occur somewhere on the planet every year. Voilà. So he finally has his perfect out" for rationalizing his predicted catastrophes.

If a storm compels residents to flee an area with decent economic, moral, and social circumstances, and without rampant, notorious corrup-tion, (and with above-sea-level elevations), those residents are more likely to return and therefore less likely to be "climate refugees." The politicians and original city planners of New Orleans may present its denizens with all sorts of reasons not to come back, but now, in Brown's view, all of those sins are washed away, and there is only one cause: cli-mate change. The opportunities for parody are endless. Little Lester with a ball glove and broken window, telling Mom: *global warming*. Pulled over by the cops? *CO_2 concentrations*.

Topping this episode is the unintentionally hilarious doomsday-pro-claiming annual *State of the World* volumes of Brown's followers. The pinnacle of his achievement was one of the reports predicting nothing short of the eradication of Switzerland. Ever-rising temperatures causing melting and destabilization of the massive winter mountain snow cover (and glaciers) would prompt vast avalanches which would deforest the mountainsides, followed later by the increasing melting snow and water runoff creating vast mud and rock slides racing down (tree-free) mountains at great speeds and totally annihilating picturesque Alpine villages and lower towns, *etc.*

This is a wonderful scenario for fundraising and selling books, getting on TV and radio, and generally doing things that environmentalist elites like to do. Not subscribing to any Swiss newspapers I'm not sure if it has come true. Someone might check.

Don't bother telling Brown, however, as like Ehrlich he seems never to be embarrassed over the rate at which his predictions actually come true. Keeping Ehrlich's flame alive and sufficiently erratic, World-Watch manages to now complain about obesity, as well as economic growth, in its 2006–2007 report.[17] On the rare occasion when challenged, the alarmist response distills to something like, "Well that's your fault as a reader. I'm a scientist. I never make predictions. I just sketched a few possible scenarios. Which could very well happen someday." Watch this space. Subscribe to next year's issue!

> ## False Prophecies
>
> This "nutritional disaster seems likely to overtake humanity in the 1970s (or, at the latest, the 1980s).... A situation has been created that could lead to a billion or more people starving to death."
>
> **Ehrlich** in *The End of Affluence*, 1974

Global warming alarmism: Who's the greatest?

Bill Clinton has taken a fancy to the description of "global warming" coined by British prime minister Tony Blair's science advisor Sir David

King as *"the greatest threat facing mankind," "worse than terrorism."* This sentiment is echoed by environmentalist activists and journalists, though Blair himself attempts to temper Sir King's unhinged assessment to *"the greatest* environmental *threat."*

Gore advisor and father of global warming Dr. James Hansen now warns that it is nearly too late to act,[18] while others claim that such an assessment is too rosy.[19] The obvious question for the greens is why they continue to push the timid Kyoto proposal, given that even if one accepts each and every alarmist assumption incorporated into the world's most advanced climate model, Kyoto would only avoid an undetectable warming of 0.07 degrees Celsius by 2050.[20] We know that the greens and their media pals even see this as a very, *very* small amount. We know this because they hyped the summer of 2006 as having *just missed* being the warmest ever—at 0.23° C warmer than 1936.[21]

Sometimes, despite themselves, these greens are truly just ever so helpful to our efforts aimed at keeping things in perspective.

Bjorn Lomborg emphasizes how the greens' climate proposal involves disregarding all other concerns, in pursuit of a remedy they admit will be grossly insufficient and indeed climatically meaningless. Still, this "remedy" takes priority over global sanitation, drinking water, and AIDS treatment—all of which could be provided at one-tenth the cost of one year of Kyoto. One can no doubt clearly see from this calculation how it's the Kyoto *skeptics* who are heartless.

When confronted with the inadequacy of their answers to global warming, greens respond that Kyoto is merely *"the first of thirty steps."*[22] Actually, Kyoto is the only thing that is on the table because, while having no detectable influence on atmospheric CO_2 concentrations or on temperature, it will have a huge downward effect on the planet's economic health. It was deemed to be as large a pill of energy suppression and wealth transfer that rich countries would presumably swallow. But if global warming is "*the greatest*" threat—and an imminent one—how can they offer such a do-nothing proposal in the name of incrementalism? The obvious answer is that proposing restrictions on energy use as dramatic they claim necessary (and clearly desire) would turn off most policymakers and regular people and possibly prompt sincere scientific debate over what such an expenditure would yield—which is the last thing they want.

If the greens believe what they say they believe, then Kyoto would be an immoral proposal given its impotence, due to lack of "political will," to ward off an apocalyptic climate change. The fact that it is their only proposal brings their sincerity and good will into doubt.

Being an environmentalist, however, not only means never having to say you're sorry (those millions of dead Africans killed by malaria when DDT was taken away from them might deserve an apology), it also means never having to prioritize. Upon inspection, that is because the environmentalists, like Al Gore in the words of one DNC memo discussing the political vulnerabilities of his environmental froth, have no sense of proportion. You may have noticed, for example, a proliferation over the years of "*greatest threats.*" When it comes to "climate change," however, the greens tie themselves in knots over the list of greatest perils.

Of course, environmentalists are not alone in playing this game, but the others enter the stadium once the greens design it. For example, the greens are happy to exempt from Kyoto some rather industrialized countries such as China, India, Brazil, and South Korea—countries who any year now will take over as the dominant "greenhouse gas" emitters. These countries, none of which will change their economies or energy use under Kyoto, agree that, yes, global warming is indeed the greatest threat facing mankind, so great in fact that *others* should do something about it. It's pretty handy, that this "something" will hamstring the richest countries, compelling them to offshore more manufacturing—to these very same uncovered nations.

Nuke the...nukes

A sincere response to "the greatest threat" should include all available means for reducing the alleged threat—greenhouse gas emissions in this case. Policymaking is nothing if not the art of prioritization. Therefore, even accepting the more modest baseline of "greatest *environmental* threat," it is hypocritical that Kyotophile greens almost unanimously still oppose nuclear power. The Kyoto Protocol, in fact, does not permit parties to gain

"My anxiety attacks began two summers ago. They were mild at first, a low-level unease. But over a period of months they grew steadily worse, morphing into full-fledged fits of panic. I was a wreck. The sight of an idling car, heat-trapping carbon dioxide spewing from its tailpipe, would send me into an hours-long panic, complete with shaking, the sweats, and staring off into space while others conversed around me. The same thing happened on overly warm days, like those 60-degree ones here in the Big Apple last January. ... I had come down with a severe case of eco-anxiety—a chronic fear of the environmental future.

My condition only got worse. ... I'd skip the elevator and walk the eight flights of stairs to my apartment. At night, I lay awake worrying about which of the myriad climate-related disasters scientists are predicting would come first—flood, famine, heat wave, drought ...

Nothing I did to curtail my anxiety helped—not talking to my shrink, not switching my apartment to a greenhouse gas-free electricity supplier, not handing out cards to idling motorists telling them how much pollution they could prevent simply by turning off their engines. My girlfriend started screening the newspaper for me, like some Soviet censor, snipping away alarming news. But even with her intervention, I felt alone. Riding high in their SUVs, few people around me seemed concerned about the changing climate."

Liz Galst, "Global Worrying: 'The environment is in peril and anxiety disorders are on the rise.' What's the connection?" *Plenty* magazine, August/September 2006

GHG "credits" for reducing their emissions by using nuclear power.[23] The greens openly celebrated in 2003 that they had ensured that nuclear power would *not* be one of the means to reducing CO_2 under Kyoto.

But even this claim is untrue: many countries, such as France, only stand the slightest hope of complying with their Kyoto promise because of their extensive use of nuclear power (for up to 80 percent of their electricity production). What the greens, and Kyoto, did was ensure that no credit is given to countries for providing *energy poor, Kyoto-exempt* countries with nuclear power that would of course save millions from drudgery and death. This possibly explains the greens' glee.

The claims and the treaty are inescapably insincere. France pushed for the nuclear power ban even despite its own reliance on nuclear power today. It's a muddled morality play. They've got theirs, and that's plenty good.

Killer...trees?

Nuclear power is not the only threat that the alarmists fear more than "the greatest threat." Consider dams, for example, which are impermissible to most greens as being a "renewable" energy source (hydropower),

False Prophecies

"The threat of a new ice age must now stand alongside nuclear war as a likely source of wholesale death and misery for mankind."

Nigel Calder, *International Wildlife*, June 1975

"The continued rapid cooling of the earth since WWII is in accord with the increase in global air pollution associated with industrialisation, mechanisation, urbanisation and exploding population."

Reid Bryson, "Global Ecology; Readings towards a Rational Strategy for Man," 1971

"The rapid cooling of the earth since World War II is also in accord with the increased air pollution associated with industrialization and an exploding population."

Reid Bryson, "Environmental Roulette," 1971

> "From a social/political/historical perspective climate alarmism is increasingly reminiscent of nuclear freeze agitation in the early 1980s. Jonathan Schell's surprise best-seller, *The Fate of the Earth*, could be reissued today with the words 'nuclear weapons' swapped out for 'global warming,' especially the overlay of pessimistic philosophy. Not even the ability to travel to other solar systems or galaxies offered Schell any hope, because 'wherever human beings went, there also would go the knowledge of how to build nuclear weapons, and, with it, the peril of extinction.' Today, we would take greenhouse gas knowledge with us to any other planet, threatening Mars, for example, with melting ice caps. Oh, wait, Mars is doing that on its own...."
>
> **Steven Hayward** of the American Enterprise Institute (in an email)

because by disrupting the flow of water they are mean to fish. Nature didn't intend things this way. And nature, of course, is never wrong.

The litany of "greater than the greatest" threat is ever-expanding. It entered the absurd years ago when Kyoto negotiators first recorded their fear of trees. In November 2000, Europe sought to gut Kyoto's plain language in Article 3, which states that CO_2 "sinks...shall be used to meet the commitments under this Article of each Party included in Annex I." Sinks are forestry that absorb carbon dioxide to produce oxygen through photosynthesis and other land-use practices capturing or not releasing GHGs. EU negotiators, facing a desperate Gore team (which was keeping an eye on the ongoing Florida ballot recount), seemed gripped by a fear that the U.S. might suddenly decide to reforest instead of simply shutting down.

UN and EU eco-preachers proclaimed that to accept such unchecked use of trees to absorb GHGs would be to "destroy the environmental integrity of the agreement." This, as you will see in Part IV, was too much even for the Clinton-Gore team. They walked, and abandoned the treaty and all its gamesmanship.

The irrational fear of trees as a threat greater than global warming continues to unfold. The big fight at the December 2003 "COP-9" Kyoto negotiation in Milan was again over how aggressively to accept arboreal atmospheric ardor. This time, the demand was for a limit on what *kind* of trees should be permitted to soak up the CO_2, that menacing precursor of the dreaded photosynthesis.

"Why, what if 'Franken-trees' abound?" the greens asked. "Certainly worse than global warming?" Reason, such as it exists in this context, prevailed and the campaign to prohibit countries gaining credit for GHGs absorbed by "genetically modified" trees failed.

Not to accept defeat (or much else) gracefully, the greens came back with a vengeance at the 2004 "COP-10" in Buenos Aires. "Genetically modified trees must be banned from the Kyoto Protocol!" screamed a joint press release of the Friends of the Earth International (FOE-I) and the World Rainforest Movement (WRM). The "grave decision" in Milan was a "dangerous outcome," for two reasons: one, these greens warned of the "negative social . . . impacts" of allowing non-multicultural trees to proliferate. Seriously. These "monoculture trees" were bad enough, say WRM[24] and FOE-I. (A monoculture tree is a tree growing solely among its own kind.[25] English-speaking peoples might call them "planted" or "farmed" trees . . . *shudder*.) And two, by implication, allowing such arboreal isolation, which is not how Mother Nature arranged *her* forests, is also

> "The fundamental problem, as I see it, is that environmental groups are too often alarmists. They have an awful track record, so they've lost credibility with the public . . . . I'm now skeptical of the "I Have a Nightmare" speeches . . . . Environmental alarms have been screeching for so long that, like car alarms, they are now just irritating background noise."
>
> —*New York Times* columnist and reliable green **Nicholas Kristof**, March 12, 2005

quite clearly a greater threat than climate change. The madness multiplies when we allow such trees to be genetically modified to resist weather and pests. Therefore, if any aspect of a tree's existence is the result of some deliberate choice by a human, that tree outranks global warming on the list of threats to mankind.

Even in the hyper-hyperbolic realm of environmental alarmism, such abandonment of perspective ought to stand out as a dangerous practice itself.

One almost has to feel bad for our green friends, however. After having conceded, again, the practice of allowing trees (some of them, anyway) to soak up the CO_2 that they claim is so deadly, things only got worse for our increasingly confused tree-huggers. In January 2006, research from the Max Planck Institute published in the oft-alarmist *Nature* magazine[26] established that trees and other foliage actually emit enormous quantities of methane, a GHG *thirty times more powerful than CO_2.*[27] In fact, it seems that plants apparently produce 10 to 30 percent of the total annual methane volume, suddenly revised upward. (Oh yeah, this, too, overturned years-old "consensus" on the sources and volume of methane. So much—again—for the science being "settled.")

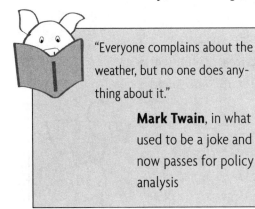

"Everyone complains about the weather, but no one does anything about it."

Mark Twain, in what used to be a joke and now passes for policy analysis

As Zeeya Merali elegantly put it in *New Scientist,* "The lungs of the planet are belching methane. It's not just farting cows and belching sheep that spew out methane. Living plants have been disgorging millions of tonnes of the potent greenhouse gas into the atmosphere every year—without anybody noticing."[28] Farting trees, *that* we'd notice.

Global warming is the greatest threat facing mankind. Gee, but those "solutions" are scary too. Hydroelectric power (dams) are mean to fish, nuclear emits nothing and so must be stopped, the trees are after us, flatulent or not, and so on. Biomass (the vaunted woodchips and switchgrass with which President Bush plans to fuel the U.S. economy) is suddenly opposed as a major substitute for "fossil fuels" by the European Union and, of course, Lester Brown for the respective reasons of possibly threatening biodiversity and unacceptably supplanting food resources. Add two more threats to the list.

If this is beginning to sound ridiculous, well, then you're paying far too much attention for the greens' tastes.

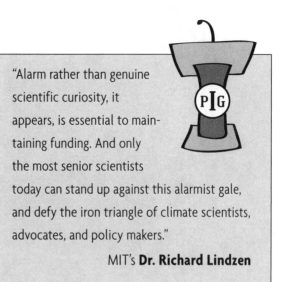

"Alarm rather than genuine scientific curiosity, it appears, is essential to maintaining funding. And only the most senior scientists today can stand up against this alarmist gale, and defy the iron triangle of climate scientists, advocates, and policy makers."

MIT's **Dr. Richard Lindzen**

Part II

✴✴✴✴✴✴✴

GLOBAL WARMING:
THE CONVENIENT LIES

Top Ten "Global Warming" Myths:

Myth 10. It's hot in here! When comedian Henny Youngman was asked, "How's your wife?" he would respond, "Compared to what?" A more critical media would ask the same question about the temperature. Present temperatures are warm if you compare today to, say, the 1970s, or to the Little Ice Age (about 1200 A.D. to the end of the nineteenth century), or to thousands of years ago. Select many other baselines, for example, compared to the 1930s, or 1000 A.D.—or 1998—and it is presently *cool*. Cooling does paint a far more frightening picture given that another ice age would be truly catastrophic, while throughout history warming periods have always ushered in prosperity. Maybe that's why the greens tried "global cooling" first.

Myth 9. The 1990s were the hottest decade on record: Targeting the intellectually lazy and easily frightened, this claim ignores numerous obvious factors. Of course, "on record" means "since we developed reliable temperature records," which generally means a very short period. Also, the National Academy of Sciences debunked this claim in 2006. Further, previously rural measuring stations register warmer temperatures after decades of "sprawl" (growth). Measurement of "global mean temperatures" also appear to have been affected when hundreds of weather stations (many in the Soviet Union's Arctic) went offline at the beginning of the decade.

Myth 8. The science is settled; CO_2 causes global warming: Historically, atmospheric CO_2 typically increases *after* warming begins, not *before*. The most common effort to dodge an actual debate over the causes of global warming is to claim that we had the debate, it is over, and there is "consensus" (about *what* is typically left unspoken). If it is really settled, why don't the scientists forgo the $5 billion in taxpayer money they get every year to research climate? What scientists *do* agree on is little and says nothing about Manmade global warming. Namely, they agree that (1) global average temperature is probably about 0.6° Celsius—or 1° Fahrenheit—higher than a century ago; (2) atmospheric levels of carbon dioxide (CO_2) have risen by about 30 percent over the past two hundred years; and (3) CO_2 is a greenhouse gas, and increased greenhouse gases

continued on page 63

continued from page 62

should have a warming effect, all else being equal (which it demonstrably is not). Regardless, "consensus" is the stuff of politics. It means ending debate in order to "move on." Stifling debate is inherently anti-scientific.

Myth 7. Climate was stable until Man came along: Swallowing this whopper requires burning every basic history and science text, just like "witches" were burned in retaliation for changing climates in ages past. The poster child for this concept is a graph that looks like a hockey stick, which has now been disgraced and airbrushed from the UN's alarmist repertoire.

Myth 6. The glaciers are melting! As good fortune has it, frozen things do in fact melt or at least recede after cooling periods mercifully end. The glacial retreat we read about is selective, however. Glaciers are also advancing all over, including lonely glaciers nearby their more popular retreating neighbors. If glaciers retreating were proof of global warming then glaciers advancing are evidence of global cooling; they cannot both be true and in fact neither is. Also, retreat often seems to be unrelated to warming, for example, the snow cap on Mount Kilimanjaro is receding—despite decades of cooling in Kenya—due to regional land use and atmospheric moisture changes.

Myth 5. Climate change is raising the sea levels: Sea levels rise during interglacial periods such as that in which we (happily) find ourselves. Al Gore extrapolates and prophecies Manhattan underwater. However, even the distorted United Nations International Panel on Climate Change refutes such breathless claims, finding no statistically significant change in the rate of increase over the past century. In other words, sea levels have steadily and slowly been rising for a very long time, and industry, rising temperatures, and increased CO_2 have not noticeably affected that rate. Small island nations seeking welfare and asylum for their citizens in, e.g., socially generous New Zealand and Australia, have no sea level rise at all and in some cases see instead a drop. These societies' real problem is typically that they have made a mess of their own situation. One notably cranky archipelago nation is even spending lavishly to lobby the European Union for development money to build beachfront hotel resorts, at the same time it shrieks about a watery and imminent grave.

continued on page 64

continued from page 63

Myth 4. Climate change is the greatest threat to the world's poor: Climate, or more accurately *weather*, remains one of the greatest challenges to the poor. Climate *change* adds nothing to that calculus, however. Climate and weather patterns have always changed, as they always will. Man has always best dealt with this through adaptation and technological advance—and most poorly through superstitious casting of blame (again, remember the witches). The most advanced, or wealthiest, societies have always adapted best. One would prefer to face a similar storm in Florida than Bangladesh. Institutions, infrastructure, and access to energy are keys to dealing with an ever-changing climate, not rationing energy use.

Myth 3. "Global warming" means more frequent, more severe storms: Here again even the UN's climate change panel doesn't support this. Storms are cyclical and, that said, are *not* more frequent or more severe than in the past. Luckily for Al Gore, reporters typically have little use for actual data.

Myth 2. "Global warming" proposals are about the environment: Only if this means that they would make things worse, given that "wealthier is healthier, *and* cleaner." Even accepting every underlying economic and alarmist environmentalist assumption, no one dares say that the expensive Kyoto Protocol would detectably impact climate. Imagine how expensive a pact must be—in both financial and human costs—to so severely ration energy use as the greens demand. Instead, proponents candidly admit desires to control others' lifestyles; supportive industries hope to make millions off the deal; Europe's environment commissioner admitted that Kyoto is "about competition, about leveling the playing field for big businesses worldwide" (i.e., bailing them out).

Myth 1. The U.S. is going it alone on Kyoto and "global warming": Nonsense. The U.S. rejects the Kyoto Protocol's energy rationing scheme, along with 155 other countries, representing most of the world's population, economic activity, and projected future growth. Kyoto is a European treaty with one dozen others, *none* of whom are in fact presently reducing their emissions. Similarly, claims that President Bush has derailed some Clinton-Gore effort to ratify Kyoto are false on every front.

Chapter Four

✳ ✳ ✳ ✳ ✳ ✳ ✳

GLOBAL WARMING 101
NOT MANMADE, CATASTROPHIC, NOR GLOBAL

Open the paper, turn on the evening news, listen to Congress on C-SPAN, step into a public school classroom, or go to a corporate seminar, and you'll get the same story:

By driving too much, using too much power, and relying too much on fossil fuels, Man is causing global warming that will be disastrous to the planet. Global warming will cause glaciers to melt and sea levels to rise.

The debate is over, they tell us. The science is proven. We have *consensus*. Anyone who doubts us is either blind or (more likely) dishonest. Skeptics are usually in the pay of the oil companies—the very companies who are "polluting" the air with dreaded carbon dioxide.

We must act now, they implore us. Thankfully real solutions are at hand. Responsible lawmakers from both parties have put forward legislation that will tackle global warming. *Even* corporate America is on board. The rest of the planet has started on their way to the solution with the Kyoto Protocol, but George W. Bush, the Texas oilman, pulled us out of the treaty. We are alone in the world, as Europe is cutting its greenhouse gases.

This story, like any good myth, is useful for those who proffer it, but it has little grounding in facts.

The main hole in the "settled" theory of catastrophic Manmade global warming is that it is not catastrophic, Manmade, nor global.

Guess what?

❊ Climate is always changing.

❊ The sun's activity correlates more closely to global temperatures than CO_2 levels do.

❊ Global warming likely would have a net benefit…as past warmings always have.

❊ Green "solutions" would be draconian and ineffective (unless their real goal is controlling economic and population growth).

Yes, on average, the planet is getting warmer. This warming seems to be mostly at night, in the winter, and at the North Pole. In fact, the Southern Hemisphere as a whole seems not to be experiencing any statistically significant warming.

The current warming is not unprecedented. Climate always fluctuates. We have just emerged from something called the Little Ice Age, and so it's no wonder things are relatively warm. Evidence suggests it is currently colder than it was during the well-established Medieval Warm Period. To raise the global warming alarm, some advocates have cleverly—though bluntly—tried to erase past climate fluctuations from the history books.

We are very uncertain about the extent of the warming. It is impossible to actually take the temperature of the whole planet. Historical temperature data is tough to discern. Most claims are based on "proxy" measurements such as tree rings, ice-core samples, and knowledge of what crops thrived when and where.

Yes, carbon dioxide acts as a greenhouse gas, absorbing radiation and retaining heat, thus making the planet habitable. Yes, burning coal, oil, and natural gas gives off CO_2. But to what extent is human activity responsible for the current warming? Probably very little. Many factors, especially the volatile sun, can contribute to temperature change. Even more factors contribute to greenhouse gas concentrations. Greenhouse gases have always been in our atmosphere, sometimes in far greater quantities than today, and are in fact a condition for human life—without any greenhouse effect, our surface climate would be comparable to Mars's.

We cannot even be sure the Earth's warming is a bad thing. Plants appreciate warmer temperature (as well as higher CO_2 concentrations). North Dakotans and Russians do, too, and most of the warming is happening in colder climes. We know the residents of Greenland prospered during the Medieval Warm Period. Then, many left.

But doesn't Arctic warming mean glaciers melting, polar bears drowning, and Manhattan going underwater? That's a lot of hyperbole, too. For every shrinking glacier there is a growing one—but the growing ones get much less attention.

These ambiguities and uncertainties all illustrate how popular claims of *consensus* on the causes, extent, and effects of global warming are misleading. The dishonesty and bully tactics employed to preserve the appearance of consensus are startling.

Those "socially responsible" corporations signing on to the global warming crusade are generally clever capitalists looking to make a buck off some government mandate, trading scheme, subsidy, regulation, or other favor. Enron was the ringleader in this racket.

Which leads us to the "solutions" to global warming, which manage to be simultaneously impotent to "stop global warming" but very potent in hobbling the economy. The Kyoto Protocol would drive up prices for all families, rapidly increase government (or UN) control, dramatically limit our ability to use energy, but would still not even prevent one-tenth of one degree of warming over the next fifty years.

Yes, there's a lot they're not telling you. Here's a starter course.

> **"WASHINGTON, D.C., September 15, 2000 (ENS)—**Parallel to the landmark lawsuits that have forced change upon the tobacco industry, one of the world's largest environmental groups today announced it may take legal action against industrialized countries and private industries that attempt to block the implementation of the Kyoto Protocol on global warming."
>
> "Friends of the Earth Considers Legal Action to Curb Global Warming," Environmental News Service, September 15–16, 2000

What's causing the warming?

The climate is always changing. Different parts of the planet are always getting colder or warmer, wetter or drier. Many things can cause this climate change. The sun has cycles, sometimes producing more energy, and sometimes producing less. The Earth's wobble and eccentric orbit mean that different parts of the planet will be exposed to varying amounts of heat over different periods. If more snow or land is exposed, more heat might be reflected. If more water is exposed, more heat will be absorbed. If the sky gets darkened by dust—caused by a volcano, a meteor, or pollution—it can make the planet colder. Land-use changes, Manmade or otherwise, greatly impact local climate. Finally, there is the most famous (but still one of only many) factor in temperature: greenhouse gases.

"Greenhouse gases" are gases that principally occur naturally. Carbon dioxide is one greenhouse gas. We make CO_2 when we breathe out. Plants release CO_2 and other GHGs when they die. Oceans store and release enormous quantities of CO_2. Nitrous oxides are greenhouse gases produced in soils by microbial processes. Methane is another. It comes from decaying plants, seeps from swamps, bogs, rice paddies, and leaks out the front and back ends of masticating animals.[1] By allowing sunlight to enter our atmosphere freely but then absorbing and otherwise trapping infrared solar radiation (heat), these gases form a protective blanket sustaining life; without them, Earth would be uninhabitable, as our atmosphere would be, for all purposes, equivalent to that of Mars.

Humans add to the greenhouse gas concentration by not just by exhaling but by harvesting plants, and releasing methane, typically after a meal of Mexican food. (*Measure that, EPA!*) But we also create greenhouse gases by the processes through which we generate or release energy—for our homes, our factories, and our cars—all processes that involve hydrocarbons.

Hydrocarbons, which include petroleum, coal, and natural gas, consist largely of hydrogen and carbon atoms. The bonds in these hydrocarbon molecules are very strong, and so breaking the bonds releases a good bit of energy. They are easily combusted, and therefore make great fuels.

To release the energy, we burn them—or *oxidize* them—and then use the freed energy to keep our houses warm, our refrigerators humming, our cars moving, and our internet servers serving. The coal or oil being burned typically possesses impurities, which can go into the air as pollution. If the hydrocarbon fuel is incompletely burned, it can give off poisonous carbon monoxide. Ideally, hydrocarbons are transformed entirely into energy and the odorless gas carbon dioxide. (The distinction between CO—poisonous carbon monoxide—and CO_2—benign carbon dioxide—is one lost on my hate-mailers who urge me to asphyxiate myself with the latter.)

As such, CO_2 is not a *byproduct* or *pollutant* but an intended result of energy production. The more efficiently one combusts a hydrocarbon, the more CO_2 one produces. This is one reason why advocates of "energy efficiency" as a global warming solution haven't quite perfected their argument.

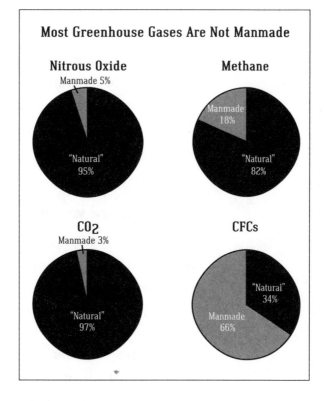

CO_2 was previously most infamous for its cruel imposition of the vicious cycle of photosynthesis upon our floral friends, forcing them to produce oxygen which we fauna then selfishly inhale, only to heartlessly exhale more CO_2.

While SUVs and power plants garner the most media and environmentalist attention, combustion emissions contribute about 2 percent of greenhouse gases currently keeping our atmosphere habitable. This bears repeating: of all the factors causing climate change, Manmade greenhouse gases are a tiny fraction of *one* factor.

Most greenhouse gases are produced by "natural" processes. Still, "greenhouse warming" theory vows that man's marginal contribution will tip the atmospheric system into some disequilibrium producing delirious climate change (the "tipping point" claim, which assumes, again, that climate was once stable).

But Man's emissions are chemically indistinguishable from Nature's own (assuming, as do the greens and Kyoto, that Man is not part of nature). Earth itself varies wildly in terms of relative volume of GHGs it produces or releases each year.[2] Somehow, by this thinking, the planet treats *Man's* contribution to greenhouse gases differently. So the argument goes.

How much are we warming?

Warming is happening, but it is slight, it is relative, and it is not "global" in that it is not warming everywhere (for example, the Southern Hemisphere).

To keep alive the story of catastrophic Manmade global warming, the greens managed to rewrite not just ninth grade biology such that carbon dioxide is now a "pollutant," at least rhetorically,[3] but also history in that a mythical millennium of climate stability was interrupted by the Industrial Revolution, requiring airbrushing the record of the Little Ice Age and Medieval Climate Optimum (or "warming," so named due to the advances in technology and human life spans occurring as a result of the warming).

Alarmists proclaim the 1990s as the "hottest decade," pointing to the infamous "Hockey Stick" graph of temperature reconstructions melded onto actual temperature measurements. It turns out that the '90s not only fail to live up to the "hottest" title, but coincided with the closure of hundreds of measuring stations (including many in the former Soviet Union as their priorities turned to more pressing domestic matters such as collapse of an empire). If you shut down measuring stations in the cold parts of the world, your average global temperatures will go up. It turns out that the 1990s' temperature increases track nicely with these closures. At the very least, one should be wary comparing post-1990 temperature averages with data from before the massive shut-down of stations. Yet not one journalist can be bothered with this concern.

While anecdotal observations serve as the basis for prominent climate scare stories in the media, the principal basis for alarmism is computer climate model projections. These, as with any model, can be designed to produce whatever outcome is desired.

Reconciling computer model projections with real-world experience (observations over the past three decades of a 0.17-degree C rise per decade) suggests that we might expect a warming of about 1.7 degrees C over the next century. [4] Further, all but two among dozens of climate models predict linear warming (steady), not exponential (skyrocketing, as in Al Gore's celluloid fable). Remember, predicting that the planet will get warmer says nothing at all about what—if any—contribution Man might have, but this does raise the question of *why the hysteria*?

In a crushing blow to the media, it appears that things are turning out even milder than that. In May 2006, the National Oceanic and Atmospheric Administration issued a report, co-authored by Dr. John Christy, that worked out even more the differences in temperature trends between surface and satellite measurements (a major controversy given that the atmosphere was supposed to warm first and worst, but couldn't seem to keep up with the surface, which was being layered with more heat-absorbing concrete all the time). The report found that "global-average temperature increased at a rate of about 0.12 degrees C per decade since 1958, and about 0.16 degrees C per decade since 1979. In the tropics, temperature increased at about 0.11 degrees C per decade since 1958, and about 0.13 degrees C per decade since 1979." This is far less than the models generally predict, and, as such, Christy noted (reported, surprisingly, by the *Washington Post*), the Earth is not heating up rapidly. [5]

Further, to claim "global warming" with any degree of accuracy one must be referring to an increase in measured global mean surface temperature—a quantity that has never actually been measured. Surface temperature is not measured globally but rather haphazardly, wherever measuring stations have been placed. As noted, coincidentally enough the number of measuring stations changed drastically immediately prior to the "hottest decade on record," and stations in poorer countries are maintained differently compared to those in wealthier countries. This is why one leading climate scientist says that the "global mean surface temperature" means as much to him as would the global mean telephone number.

Note also that this does not make today's average temperatures warm by historic standards. Considering that it is only warm right now if you deliberately choose as your baseline a year colder than today, "global warming" has been occurring since the six- to seven-hundred-year cooling period known as the Little Ice Age ended—to the tune of about one degree Fahrenheit in the past one hundred-plus years. That's what all of the fuss is about. A degree is supposedly responsible for all of the tales of woe, despite that it's been warmer, and cooler, in the past. Further, it's not very meaningful to say that the world warms after cooling periods end. Man is likely responsible for at least some fraction of the warming, though whether that contribution is detectable is unknown. For those who pay attention to short periods, it warmed until the mid-1940s, cooled until the late 1970s, and then warmed again.

The warming of the Earth's surface that many scientists associate with Manmade greenhouse gas emissions is actually distributed in the least catastrophic—and most *beneficial*—fashion possible. During the recent, slight warming trend, it is the nighttime and winter temperatures that have seen any notable increase—and tilted toward the northern latitudes, if not so much as predicted. In fact, winter, polar, and nighttime warming has accounted for nearly all of the counted warming, increasing twice as much as the daylight or summer ("maximum") temperatures. This translates to longer growing seasons and warmer nights, which foster plant growth and agricultural productivity by reinforcing the fertilization effect of carbon dioxide. *Oh, the humanity.*

Yet this does not mean that the warming trend of the past three decades will continue at its current pace, much less at a catastrophic pace. (It also may not be a bad thing on net.) Consider my infant son. He grew nearly a foot in his first twelve months on this planet. Shall we assume that in mere years he will be terrorizing skyscrapers as any giant should? Of course not. Like the impact of carbon (carbohydrates) on an infant's growth pattern, the global warming impact of the GHGs that Man adds to nature's mix is logarithmic, not linear. That is, my son will not in fact grow to be forty feet tall and weight a ton; he will grow steadily then cease doing so (as with climate, certain regional exceptions will arise, like difficulty buttoning his bell bottoms). People and the atmosphere are of course different systems, but the analog reminds us that if we double the amount of CO_2 in the air, say, by adding x CO_2 to an atmosphere already containing x CO_2, and get a warming effect of y degrees, we don't get another y degrees of warming just by adding another x CO_2. You would need to go up to $4x$ to do that. Thus, warming is logarithmic.

In short, the evidence cited for catastrophic Manmade global warming does not credibly demonstrate that Man is capable of causing the sort of greenhouse calamity promised by the alarmists to justify their "bold" "solutions." Notably, that "cure" is the same cure that has been offered over decades for any number of ills both real and not-so-real, including global cooling: drastic cuts in energy use, combined with the environmentalist community's long-standing goal of far, far fewer people occupying the Earth and using its resources.

How destructive is this warming?

An increase of 0.12, 0.17, whatever. Though they insist the warming will be dramatic, when pressed the alarmists promise that a temperature rise of even 1.2–1.7 degrees C—over a century—will be calamitous. This raises some questions:

Don't most flora and fauna live in warmer areas for the very good reason that they survive better there, actually, the same reason that 90 percent of Canadians are huddled near their southern border? Hasn't it been warmer than that in Man's earthly experience? Has the planet—and life on it—adapted and thrived after such temperature increases in the past? Hasn't the Earth experienced even the projected *rate* of warming before? In fact, doesn't all planetary life experience major temperature swings between midnight and noon nearly every day, and much more throughout the seasons of the year, clearly without mass extinction or catastrophe?

The answer to all of these is "yes."

How about storms, of which we hear so much? The Atlantic Basin and Northeast Pacific are the only two areas that have actually faced a statistically meaningful change in severe storm activity in recent years. You've heard about the former, but not the latter. This is no doubt purely coincidental with the fact that the Northeast Pacific has seen a *decrease* in activity. The increase in Atlantic storms was long-predicted as a resumption of a well-known forty- to fifty-year cycle. While not completely ignored, this was largely lost in the news coverage; it's somewhere with all the stories about the very quiet 2006 hurricane season.

Finally, we need to ask whether warmer is necessarily worse.

Russian president Vladimir Putin made waves when rhetorically asking why a cold country such as Russia would fear a couple of degrees of warming. (Indeed, the Russian Academy of Sciences last November warned about the ice age returning.) Given geographic retirement trends, Putin does seem to have a point. Cold is not only not pleasant, but it kills like heat rarely can, as is statistically borne out *whatever* one's baseline. For example, the UK Department of Health calculates that, if the southern UK warmed by 3° C by the 2050s, as some claim it might, 2,000 more people would die in summer heat waves each year, but 20,000 fewer people would die of cold in the winter.

Even substantial global warming would likely be beneficial to the United States. As Yale economics professor and climate expert Robert Mendelson testified to the Senate:[6]

> Climate change is likely to result in small net benefits for the United States over the next century. The primary sector that will benefit is agriculture. The large gains in this sector will more than compensate for damages expected in the coastal, energy, and water sectors, unless warming is unexpectedly severe. Forestry is also expected to enjoy small gains. Added together, the United States will likely enjoy small benefits of between $14 and $23 billion a year and will only suffer damages in the neighborhood of $13 billion if warming reaches 5C over the next century. Recent predictions of warming by 2100 suggest temperature increases of between 1.5 and 4C, suggesting that impacts are likely to be beneficial in the U.S.[7]

Greedy North Americans Using More CO_2 Than We're Producing?

"I will concede, the measurements are not perfect, they are just pretty good; they come from *Science* magazine, and they are the best numbers currently at hand—America today is apparently sinking more carbon out of the air than it is emitting into it. What's doing the sinking? In large part, regrowth of forests on land that is no longer farmed or logged, together with faster growth of existing plants and forests, which are fertilized by nitrogen oxides and carbon dioxide 'pollutants.'"

Peter Huber, "Hard Green: Saving the Environment from the Environmentalists!

The shame

So low has the discourse fallen that the European Parliament in a draft resolution blamed Hurricane Katrina on Manmade global warming, rhetorically winking at the supposed culprit (the U.S.).[8] They did so in self-parodying fashion, stating how the body "... *notes with regret that the often predicted impact of climate change has become a reality in that poor sections of society living in coastal regions bore the brunt of the hurricane.*" Of course. If it weren't for that darn climate change, poor sections of society living in coastal regions wouldn't bear the brunt of hurricanes; they would continue staring happily overhead as storms passed on to abuse the wealthier people living inland.

In fact, while many European commentators and individual politicians toed the line of actually blaming the storm on the United States, some crossed it with abandon.[9] Even accepting their premise that Man causes weather, in typical fashion this ignores that Europe's CO_2 emissions have gone up markedly since Kyoto, which cannot be said about the U.S. (all while the U.S. economy and unemployment improved markedly, which cannot be said about Europe). Were one inclined to stoop to continental-style environmental hysterics, one might therefore be tempted to say that in addition to inflicting the cruelty of its welfare and economic policies on its own people Europe caused Hurricane Katrina. Again, somehow not just historical relevance but the nastier rhetoric from our moral superiors was lost in the media coverage.

The important thing to remember is that hurricanes—like malaria, floods, and the entire "global warming" parade of horribles—happen with or without "global warming" as posited, and the "cure" of policies imposing suppressed energy use, like the Kyoto Protocol, make no one any safer, but only poorer and less able to deal with these ever-present threats.

Advancing glaciers can be found within miles of their melting brethren yet the former watch in loneliness as overheated journalists flock to the more cooperative ice. Similarly, the vaunted disappearing ice caps generally aren't disappearing. Much melting activity began at the end of the Little Ice Age and continues, often found in areas that are actually experiencing decades-long cooling. In fact, the Earth's atmospheric temperature (arguably a more relevant measure, given that anthropogenic global warming is an *atmospheric*, not a surface, theory) is not warming, like the surface (which is disproportionately influenced by development, and therefore will of course increase). Today's temperatures are about the same as in the 1930s and cooler than a thousand years ago. Someday the Vikings may be able to resume their agricultural lifestyle on Greenland.

Climate modelers will tell you they can predict cooling if that's what is desired, but for years elected activists made clear that warming was on order. With $5 billion in taxpayer dollars now at stake annually, bucking that edict would get you professional Siberia, which for some actually occurred. These massive sums are the lifeblood of research science, and so few risk rocking the boat. What prompted such enormous expenditure was a prior reliance upon twenty or so years' data to generate "consensus" panic, over "global cooling," serialized in *Newsweek* magazine. *Newsweek* remains ever-vigilant now against Manmade global warming,

"In the long run, the replacement of the precise and disciplined language of science by the misleading language of litigation and advocacy may be one of the more important sources of damage to society incurred in the current debate over global warming."

MIT Sloan Professor of Meteorology **Dr. Richard S. Lindzen** (Testimony before the U.S. Senate Committee on Environmental and Public Works, June 10, 1997)

yet so far as I can tell never particularly noted the stunning nature of how Man's fate has been reversed. Isn't *that* news?

Still, Big Green has much material to work with should it reverse course without announcement or admission, back into "global cooling," *à la* the 1980s' flip. Indications are the media would follow with hardly a blink, given that every outlet from the *Washington Post* to *Time* and even *Science* magazine flipped from cooling to warming as fast as George Orwell's Oceania flipped its historic and permanent allegiance to Eastasia in Orwell's *1984*. Remember, "Eurasia is the enemy. Eurasia has *always* been the enemy."

Green alarmism has become more breathless, more convoluted, and more well-coordinated with the establishment press as the "skeptics" continue to win on the relevant economic questions and science continues to be bipolar, so to speak (as science, frankly, should be). The global warming agenda is after all the brass ring, the mother lode, the movement's be-all-and-end-all. Victory means control over energy policy, and many individual freedoms long loathed by the greens.

Cure worse than the possible disease

That prescription itself is the greens' real goal, not remedying any particular environmental phenomenon. Control energy and you control the economy. Kyoto and its ilk seek to ration energy use. It is not an energy efficiency treaty, as some of its proponents seek to hornswoggle the public into believing. If that's what they want, they're free to draft it up.

Kyoto is rationing, plain, simple, and expressly so. Given foreseeable technologies cutting emmissions means rationing energy use, which the greens have long seen as the enemy. This is not, by the way, a pessimistic argument. Indeed, it is the greens who persistently doubt Man's innovativeness, arguing since the early twentieth century through today's climate models that Man will continue to use today's technology far into the future. Instead, it reflects that we are in the hydrocarbon age, which we entered because the greens' preferred "new" technologies like solar and windmills were not sufficient to power the Industrial Revolution. Presumably someday some technology will replace that which we have as a reliable and abundant source. But it won't be the unreliable, intermittent wind or sun. These will remain niche technologies, no matter how shrill their investors get about "global warming" to further guarantee their subsidies and mandates.

Much of the increased energy demand projected over the next three decades will be in rapidly developing countries, such as India, that will be bringing electricity to hundreds of millions of people. Until people have electricity and some access to modern transportation they can't enjoy the benefits of modern industrial civilization. Bringing power to those hundreds of millions who now must spend several hours a day gathering brush or firewood or cow dung to cook their meals will have enormous benefits far outweighing any remotely feasible, negative consequences.

This massive increase in energy use will improve people's lives, allow them to work much more productively and thereby raise their incomes, improve their health, and improve environmental quality. Certainly in India as well as other tropical and semi-tropical places, most of this

An Inconvenient Truth

"The greenhouse effect must play some role. But those who are absolutely certain that the rise in temperatures is due solely to carbon dioxide have no scientific justification. It's pure guesswork."

Henrik Svensmark, director of the Centre for Sun-Climate Research, Danish National Space Center, as quoted in the *Copenhagen Post*, October 4, 2006

energy will come from coal because they don't have much prospect for wind energy, but many of them do have lots of black gold.

The International Energy Agency's 2006 World Energy Outlook[10] agrees with other estimates of future energy demand as the world continues to develop and the poor continue to emerge into wealth. By 2030 we will require massively more energy than we use today and this will not come from windmills, solar power, or biofuels (despite great percentage gains these niche energy sources should nonetheless make).

Projected, dramatic increases in energy demand is actually the good news which, of course, the greens see as the nightmare scenario. Yet consider their preferred outcome. The world is at present energy poor, not energy rich. Starving the world's poor—or rich—of access to modern energy means starving the world's poor. Moreover, no matter how badly activists might desire to do so, the scientific community is simply not equipped to drive the debate on questions that are at heart economic or political, such as the wisdom of schemes like the Kyoto Protocol.

The "solution" to global cooling, as with warming, was to stop having babies, adopt riskier lifestyles away from which we have technologically developed, and cede national energy budgets to a supranational body prescribing each nation's ration. But the science that would support rationing energy continues to elude them.

Global warming policies to put the world on an energy diet will on the whole threaten human welfare.

Chapter Five

✳ ✳ ✳ ✳ ✳ ✳

THE "CONSENSUS" LIE
DECLARING THE DEBATE "OVER" BEFORE IT BEGAN

George Bush, the media tell us, is *anti-science*. In truth, that label is most accurately applied to the global warming crowd. Their claims of "consensus" about the causes, the extent, and the consequences of climate change—and the means they employ to preserve the appearance of this "consensus"—fairly well define the opposite of science. They are, instead, politics.

The consensus claim is a critical one for the politicians. Because their proposed "solutions" are so drastic (when it comes to lifestyle changes and government control, that is; they are toothless when it comes to affecting the climate), any doubt about the coming apocalypse would render the "solutions" politically unfeasible. Accordingly, when someone questions the hypotheses behind the global warming talk—a cornerstone of something called the *scientific method*—he is cursed as a charlatan, probably in the pay of someone who stands to profit from the destruction of the planet. If you slander everyone who questions you, maybe someday your claims will stand unquestioned.

But the consensus claim depends on discredited reports, character assassinations, and fake experts.

Guess What?

✻ Many climate experts doubt the media-proclaimed global warming alarmism

✻ Not long ago the media claimed certainty and "consensus" about global cooling

✻ Al Gore's mentor was actually a climate "skeptic"

✻ It is the Greens who seek to censor science and intimidate dissent and debate

There is "consensus" and there is truth

There is no "scientific consensus" that extreme or damaging global warming will occur or that Man is the principal or even a quantifiable determinant of climate, let alone that global warming would be a bad thing (past warmings—yes, including warmer than the present—have always been positive; dark ages have tended to coincide with cooling phases). In fact, it is difficult to identify another issue of scientific inquiry over which the debate rages more intensely.

A quick spin down Al Gore's information superhighway reveals a large number of on-line scientific debates. Two primary debate forums are the alarmist www.RealClimate.org, typically scattered with snide remarks about the heretic skeptics, and the realist www.ClimateAudit.org. Other key forums include "Still Waiting for Greenhouse,"[1] and rhetorical respondent "What's Wrong with Still Waiting for Greenhouse?"[2]

Research reveals that a few, typically quite narrow, areas of general agreement do exist regarding climate change, for example that the climate is always changing and continues to do so. Currently, on average, the planet is warming. Industrial activity is adding to the levels of "greenhouse gases" in the atmosphere. Greenhouse gases, including CO_2, can have a net warming effect, all else being equal. Beyond that, there is no consensus.

Most areas of scientific agreement are trivial and uncontroversial. One of those *not* trivial matters is the general agreement that the Kyoto Protocol will do *nothing* detectable to stop whatever warming would happen. Tom Wigley, a senior scientist at the U.S. National Center for Atmosphere Research, estimated that over the course of fifty years, the Kyoto treaty would slow global warming by .07° C, unnoticeably chipping away at the than the 2–5° C rise we are told to expect.

Even though a policy will do nothing *climatically*, politicians and reporters still call it essential on the grounds that it is "doing something" (playing to an unappeasable crowd and crippling the U.S. economy, mostly).

"Consensus" claims about the actual scientific understanding of climate either are wishful thinking cut from whole cloth, or mischaracterize the scientific research incorporated in reports by the United Nations Intergovernmental Panel on Climate Change (IPCC) or U.S. National Academy of Sciences (NAS).

For example, Dr. Julian Morris of the UK-based International Policy Network notes that "[T]he IPCC is not a scientific body: it is a consensus-oriented political body. An examination of the IPCC process [available on the IPCC's own website—www.ipcc.ch] makes it clear that the choice of authors and reviewers as well as the final review of its Reports is conducted by government officials, who may or may not be scientists. In any case, science is inherently antithetical to consensus: science is a process that involves continuously questioning and challenging what we know in order to improve our understanding of the world."

These documents only come close to claiming "consensus" in their summaries, which are written by different authors than the substance and generally mischaracterize the underlying work. The summaries, though, are typically the only part a reporter or politician's speechwriter ever reads.

MIT's Lindzen noted in a *Wall Street Journal* editorial, in the context of a National Academies report touted by alarmists as supporting their faith, that it

"The most recent survey of climate scientists, following the same methodology as a published study from 1996, found that while there had been a move towards acceptance of anthropogenic global warming, only 9.4% of respondents 'strongly agree' that climate change is mostly the result of anthropogenic sources. A similar proportion 'strongly disagree.' Furthermore, only 22.8% of respondents 'strongly agree' that the IPCC reports accurately reflect a consensus within climate science."

Professor **Dennis Bray**, GKSS Forschungszentrum, Geesthacht, Germany, submitted to *Science* on December 22, 2004, but not accepted

in fact did no such thing: "As one of eleven scientists who prepared the report, I can state that this is simply untrue....As usual, far too much public attention was paid to the hastily prepared summary rather than to the body of the report."[3] Lindzen also explained how the same phenomenon attaches to the alarmists' "bible":

> The panel was finally asked to evaluate the work of the United Nations' Intergovernmental Panel on Climate Change, focusing on the Summary for Policymakers, the only part ever read or quoted. The Summary for Policymakers, which is seen as endorsing Kyoto, is commonly presented as the consensus of thousands of the world's foremost climate scientists. Within the confines of professional courtesy, the NAS panel essentially concluded that the IPCC's Summary for Policymakers does not provide suitable guidance for the U.S. government. The full IPCC report is an admirable description of research activities in climate science, but it is not specifically directed at policy. The Summary for Policymakers is, but it is also a very different document. It represents a consensus of government representatives (many of whom are also their nations' Kyoto representatives), rather than of scientists. The resulting

"Senator James Inhofe, Chairman of the Committee on Environment and Public Works, describes global warming as 'the greatest hoax ever perpetrated on the American people' and uses McCarthy-like tactics to threaten and intimidate scientists."

James Hansen in a May 2006 draft submission to the *New York Review of Books* (even the NYRB found this to be too much and excised it)

document has a strong tendency to disguise uncertainty, and conjures up some scary scenarios for which there is no evidence.[4]

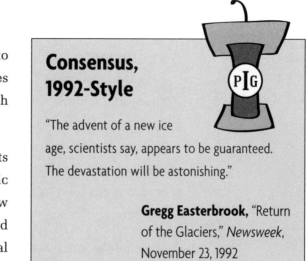

Consensus, 1992-Style

"The advent of a new ice age, scientists say, appears to be guaranteed. The devastation will be astonishing."

Gregg Easterbrook, "Return of the Glaciers," *Newsweek*, November 23, 1992

Instead of such obvious political snippets mischaracterizing underlying scientific opinion, it is more illuminating to view the work of groups whose design and membership are less inclined to political pursuits and therefore less inclined to making fantastic claims, e.g., the "Policy Statement on Climate Variability and Change" by the American Association of State Climatologists (AASC).[5] This document soberly offers points of agreement among actual scientists, and discussion thereof, including that "[p]ast climate is a useful guide to the future," "[c]limate prediction is complex with many uncertainties" and "[c]limate prediction is difficult because it involves complex, nonlinear interactions among all components of the earth's environmental system," and "[p]olicy responses to climate variability and change should be flexible and sensible."

Veteran Canadian journalist Terence Corcoran has extensively documented the travails of science under siege by alarmism. About the political push to exclude debate through shouting "consensus," he invokes actual examples of the anti-science of climate change:

> In short, under the new authoritarian science based on consensus, science doesn't matter much any more. If one scientist's 1,000-year chart showing rising global temperatures is based on bad data, it doesn't matter because we still otherwise have a consensus. If a polar bear expert says polar bears appear to be thriving, thus disproving a popular climate the-

ory, the expert and his numbers are dismissed as being outside the consensus. If studies show solar fluctuations rather than carbon emissions may be causing climate change, these are damned as relics of the old scientific method. If ice caps are not all melting, with some even getting larger, the evidence is ridiculed and condemned. We have a consensus, and this contradictory science is just noise from the skeptical fringe.[6]

Insistence on the existence of a scientific "consensus" is the product of alarmists believing that no honest person could disagree with them and that therefore anyone who disagrees is dishonest and ought to be ignored. Almost without fail, the skeptics are charged with being stooges of industry—a charge that neither addresses the skeptic's criticism or question, nor reflects the fact that much of "industry" *supports* the alarmists' agenda and often the alarmists themselves.

Naturally this has manifested itself in bizarre ways, as we shall see. One of the factors influencing Corcoran, among others, is an "open letter" sent in 2006 to the Canadian government.

The authors, sixty scientists and experts in relevant sub-disciplines, recommended the government re-open, actually, "*open*," Kyoto to debate

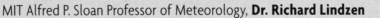

"**W**ith respect to science, the assumption behind the [alarmist] consensus is that science is the source of authority, and that authority increases with the number of scientists [who agree]. But science is not primarily a source of authority. It is a particularly effective approach of inquiry and analysis. Skepticism is essential to science—consensus is foreign."

MIT Alfred P. Sloan Professor of Meteorology, **Dr. Richard Lindzen**

on the grounds that: "when the public comes to understand that there is no 'consensus' among climate scientists about the relative importance of the various causes of global climate change, the government will be in a far better position to develop plans that reflect reality and so benefit both the environment and the economy."[7]

How dare you speak!

Presenting people who dissent from the politically determined zeitgeist a forum tends by definition to dispel the myth of consensus. This is one reason alarmists come down so hard on those outlets offering the platform. As noted, establishment journalists in the U.S. now question whether providing balance is in itself a form of bias.

In response to the Canadian letter denying consensus, a rather petulant group (dominated by environmental activists, members of the UN climate team, grant recipients, and of course government employees) wrote their own self-rebutting public appeal to the government insisting that yes the science *is too* settled, and it is time to move on from a fruitless debate (a debate that, frankly, one is hard-pressed to really remember ever happening).

Distilled, their tantrum-like reply was that *the dissent you hear does not exist as we do not sanction it, and the consensus is whatever we believe.* Keep that money and power coming!

"Consensus" outside of well-tested principles—such as the knowledge that heavy things fall—is a curious animal to find in scientific debate, to begin with. It is a far different thing to claim "consensus" agreement that Man is responsible for global warming, which will be catastrophic. To do so is abusive, and knowingly false with the obviously anti-scientific aim of stifling debate. Consensus means that everyone has settled on something. Typically in practice consensus is a "small p" political tool to produce an outcome that no one agrees with fully but everyone can live with.

"Scientific method" is something very different. It involves forming opinions through proffering and testing hypotheses. Not quite the same thing. After all, in climate science, principles and procedure are set aside—such as when the scientists Eugene Wahl and Caspar Ammann announced, by press release[8] and with subsequent fanfare, their results purporting to confirm the "hockey stick" graph discussed later in these pages. Unfortunately for them, if not for "science," their article was rejected by the *Journal of Geophysical Research Letters* (twice, as were similar comments by a David Ritson),[9] even despite its desirable conclusion. But they were able to have their argument disseminated and treated as if accepted regardless, because it satisfied journalists' template story. Yet even the collection of data through observation is subject to manipulation to fit the hypothesis—as when satellite temperature measurements were "corrected" by researchers not actually involved in the data collection. As for hypothesis testing—we will only see if the models are right decades hence. *Can't wait, no time, let's move on!*

Given this, despite their ritual shriek of "flat earther!" when confronting disagreeing heretics,[10] it is fair to point out that ancient Greek mathematician and world traveler Pythagoras was the skeptic to the *consensus* that the world was flat, and only a nut or a liar would challenge it (probably a stooge of the globe-making industry). History is full of efforts to stifle innovation by reference to the unchallengeable authority of "consensus." Galileo and Copernicus, for example, were offered opportunity to reflect on the prudence of challenging consensus.

Science requires observation—not just selectively pointing to compliant glaciers or to computer projections whose outcomes are directly dictated by the assumptions. Science requires the *testing* of hypotheses. In other words, science is skepticism, it is the practice of holding out a hypothesis for others to challenge. Compare that with consensus's "unanimity" requirement, or in its most forgiving form, majority rule.

Not only has the odious "consensus" taken root rhetorically as regards climate science, but in practice it assumes a quite vicious form. As such, it is particularly loathsome that alarmists and the rest of the global warming industry severely assail scientists or other experts with typically *ad hominem* campaigns to discredit them, and even run them out of their professional sinecure for the sin of challenging the hypothesis.

Al Gore's undercount

Al Gore states in his slide-show-turned-movie, *An Inconvenient Truth*, "There is as strong a consensus on this issue as science has ever had." Wow, considering the strength of scientific consensus about, say, the tendency of heavy things to go down, that's a pretty high standard. Gore's proof? "A survey of more than 928 scientific papers in respected

"Greenpeace co-founder and former leader [and now noted skeptic of climate alarmism and indeed modern green pressure groups, generally] Dr. Patrick Moore said the United Kingdom's Royal Society should stop playing a political blame game on global warming and retract its recent letter that smacks of a repressive and anti-intellectual attitude. 'It appears to be the policy of the Royal Society to stifle dissent and silence anyone who may have doubts about the connection between global warming and human activity,' said Dr. Moore. 'That kind of repression seems more suited to the Inquisition than to a modern, respected scientific body,' said Moore."

Newswire, September 21, 2006

Dangerous Talk

"People come to me and say: 'Stop talking like this; you're hurting the cause.'"

Dr. Robert Giegengack, geologist at the University of Pennsylvania, who has cited other causes—aside from CO_2 and greenhouse gases—for climate change in the past.

journals shows 100 percent agreement."[11] This strange if typical Gore formulation—"more than 928"?—is his money quote, and its shoddiness (unraveled below) says all one needs to know about the man and his movie.

Gore precludes that even one reputable scientist disagrees with what he says (a clever self-affirming ploy among environmentalists, in that disagreement with them inherently disqualifies one from being reputable or credible). Unfortunately, the "more than 928" papers—which the survey's author absurdly claimed represented the universe of scientific literature on the topic—actually represent less than one-tenth of the relevant scientific literature. Further, even the cherry-picked articles are in no way unanimous on the issue at hand.

Gore states (inconsistent with his other claim of "all of" the relevant research) that the author examined a "large random sample" of scientific articles. No, she did not. Intentionally or through, shall we say, inexperience, she got her search terms wrong and (we will charitably assume) presumed to be looking at all the articles when in fact the relevant scientific literature amounts to *over 11,000* articles.[12] Her research is substantively meaningless, while its use by the alarmists speaks volumes.

The author, a history instructor named Naomi Oreskes,[13] snarled that uncertainty about global warming is unwarranted "nonsense" that must now cease, because no paper in her search "refuted" her so-called consensus position on Manmade global warming. Oreskes's curious standard, then, is that she has found universal acceptance having looked hard and didn't find *refutation*.[14]

She didn't really look that hard, though. She conducted a computer search for articles in peer-reviewed scientific journals that contained the phrase "global climate change." She found 928 since 1993. In her article explaining her findings, however, she claimed to have searched for all articles using the phrase "climate change"—a search that would have yielded about 10,000 more articles.

Oreskes claimed to have reviewed these articles and proven that "there is a scientific consensus on the reality of anthropogenic climate change." She continued: "Climate scientists have repeatedly tried to make this clear. It is time for the rest of us to listen."

Oreskes claimed that *Manmade* (anthropogenic) global warming had not been questioned once in any of the relevant papers since 1993. Specifically, she claimed that "75 percent [of the studies] . . . either explicitly or implicitly accept the consensus view; 25 percent dealt with methods or paleoclimate, taking no position on current anthropogenic climate change. Remarkably, none of the papers disagreed with the consensus position." Particularly given her obvious intimation of having read the papers, her thesis is a dog's breakfast of half- and non-truths.

"Scientists have an independent obligation to respect and present the truth as they see it."

Al Gore, *An Inconvenient Truth*

* * *

"Gore's circumstantial arguments are so weak that they are pathetic. It is simply incredible that they, and his film, are commanding public attention."

Professor **Bob Carter**, Marine Geophysical Laboratory, James Cook University, Australia[16]

In reality most of the 928 papers do not even *mention* anthropogenic global warming, let alone confirm alarmism in their conclusions.[15] Her search parameters limited the universe of literature to 928 papers. Further, some papers merely *assumed* for their purposes that rising CO_2 levels from burning hydrocarbons will affect the climate, as opposed to having research findings establishing this. Most didn't present any analysis or conclusions at all about it. British social scientist Benny Peiser found that only thirteen of these articles (less than 2 percent) actually argue her purported "consensus" view. They merely mention in one context or another "global climate change" which, no one disputes, is occurring now as it always has and always will.[16] (Indeed, if one of these articles had posited "global climate *stasis*," it would be forecasting an unprecedented event in the planet's history.)

Why would articles examining something unrelated to Manmade climate change mention "global climate change?" Many scientists now throw a party-line paragraph about global warming into articles or grant proposals because it helps one gain publication and/or funding. Several researchers are on record complaining about editors requiring such obeisance. This is equal parts scandalous and illustrative about the state of science and the insidious influence of billions in taxpayer funding, so sensitive to political influences and considerations.

Even before her flawed methods came to light, Oreskes made it clear she was less a researcher than an advocate demanding that policy now follow her rhetoric. Just as she cherry-picked her search terms, Oreskes was selective when determining who among industry might be driven by financial motives to hold the positions they do.

Following the alarmist script, Oreskes humorously played this *dark and powerful forces* card. She whined in boilerplate, "some corporations whose revenues might be adversely affected by controls on carbon dioxide emissions have also alleged major uncertainties in the science," willfully ignorant to those whose revenues, thanks to clever or even cynical positioning, might be *positively* affected by such controls which they also happen to advocate: General Electric, British Petroleum, Cinergy (now part of Duke Energy), solar panel and windmill companies, and of course the granddaddy of "global warming" rent-seekers, Enron (before the, um, *unpleasantness* led them to be airbrushed from the "global warming" lobby's history).

It is noteworthy that Oreskes should be so celebrated by activists who breathlessly denounce critics as "not climate scientists" and therefore having no relevant standing to address the issues (as the greens define the term to fit their momentary needs). The same tag applies to Gore and, as she acknowledges about herself, Oreskes. Her similar lack of policy acumen proved equally impotent in dissuading her supporters from touting this risible "research" from the rooftops.

Oreskes's actual search term, "global climate change" (as opposed to the more inclusive "climate change," for which she claimed she searched) was revealed in a subsequent correction by *Science*,[17] all of which was then ignored again in Oreskes's subsequent efforts at rehabilitation.[18] Al Gore also ignores this, and all the other flaws in Oreskes's claims. At a Gore slide-show presentation I attended in January 2006, he adopted his disingenuous fallback stance, arguing, basically, that the

"Why are the opinions of scientists sought [about 'global warming'] regardless of their field of expertise? Biologists and physicians are rarely asked to endorse some theory in high energy physics. Apparently, when it comes to global warming, any scientist's agreement will do. The answer most certainly lies in politics."

Atmospheric physicist and chaired MIT Professor of Meteorology
Dr. Richard Lindzen, "Global Warming: the Origin of Consensus,"
Environmental Gore, 130

Oreskes paper merely represented a *10 percent sample* of the literature but, goodness, that unanimity certainly makes this illustrative.

Oreskes also goes so far overboard as to conclude: "This analysis shows that scientists publishing in the peer-reviewed literature agree with IPCC, the National Academy of Sciences and the public statements of their professional societies. Politicians, economists, journalists and others may have the impression of confusion, disagreement or discord among climate scientists, but that impression is incorrect."

Again, as demonstrated throughout this book and elsewhere, the IPCC and NAS by no means conclude that which Oreskes claims is "clearly expressed," in a purportedly scholarly essay no less (followed by a foot-stomping tantrum of an op-ed given prominent Sunday placement by the *Washington Post*).[19] Certainly one begins to sense an instinct among alarmists to simply *deny*, as opposed to substantively *dispute* and *argue*, that which disagrees with their faith.

This series of coincidences, all biased toward claiming alarmism and consensus, is slapstick academia and advocacy. Oreskes's screed may still be cited as the basis for an absurd celluloid tale of doom, and maintain a status among the deep-green believers, but the stain on her reputation as an academic will not soon fade. She is fortunate to have entered a field notorious for rewarding such behavior.

Had Oreskes in fact searched for papers substantively treating "climate change," she would have yielded multiples of her actual search results. Confronting the totality of the actual literature on the subject, it seems certain that she would have been persuaded against pursuing her faith-based conclusion of unanimity in pursuit of policy demands, if for no other reason than the risk of exposure. Alas, sloppiness and being too clever by half won the day. Oreskes remains at her academic perch, exposed as the Ward Churchill of climate science academic research.

Silencing dissent

Gore advisor Dr. James Hansen has dined out for years on his claim that the Bush administration has muzzled his criticism of their stance on global warming, making him the most unsuccessfully silenced critic in history. It turns out he has made this claim against one President Bush or another for nearly two decades all while maintaining a close relationship with Gore.[20] Hansen's iconic status among the media as a political victim persists despite hardly exhibiting the classic symptoms of being silenced, such as staying *off* the pages of the *New York Times* for an extended period.

Hansen's shrill cries conjure a picture of Climate Cassandras having mouths stuffed with socks and wrapped with duct tape, crammed into the trunk of an old Buick en route to the Jersey Meadowlands. This is mostly incorrect. In truth, those whose voices have been run out of the debate through one form or another hail from the more sober, "look before you leap" school, and it is the cuddly environmentalists' global warming goons who lord over an unwritten speech code, which must be enforced in order to maintain the consensus.

Consider the case of none other than Al Gore. In his late 1980s book *Earth in the Balance*, and up through his recent movie *An Inconvenient Truth,* Gore attributes his interest in "Manmade global warming" to a professor at Harvard, Roger Revelle. In these versions of their relationship, Gore comes off as a young man worshipping one of the Founding Fathers of a scientific discipline the importance of which the world has rarely seen. Gore credits Revelle with showing him that Manmade global warming is the greatest threat facing Mankind. In the true story Revelle explicitly cautioned that Gore-style alarmism was *unwarranted*. This was a truly inconvenient truth. It had to be revised.

Like many others, Revelle was a significant scientist, among the voices who raised the prospect of possible human impact on the climate, looking into the issue but by no means making a name for himself as an alarmist. But setting up a station to measure CO_2 levels did not make Revelle the first to raise the idea of human influence on climate, as Gore would have us believe; in fact, Swedish chemist Svante Arrhenius predicted in the early twentieth century that burning hydrocarbons would increase CO_2 levels that would warm the climate (which thrilled him, as the prospect thrilled most Swedes and other denizens of northern latitudes for centuries). This was quite universally accepted as was the fact that atmospheric CO_2 levels were rising. Neither Arrhenius nor Revelle were alarmists, however.

Gore and his acolytes sicced academic and legal goon squads on those who would provide witness to this truth, even at taxpayer expense. The Revelle episode is a telling one as it shines a light on the tactics Team Gore

Tricks of the Alarmist Trade

"An analysis carried out by Citizens for a Sound Economy (CSE) shows that fully 90 percent of the 'scientists' who have signed a letter frequently cited by [Clinton-Gore] administration officials as evidence of scientific consensus on global warming are not qualified to be called experts on the issue. The letter, circulated by the environmental group Ozone Action, offers the names of some 2,600 alleged experts on climate change—only one of whom is, in fact, a climatologist,' noted Patrick Burns, a global warming policy analyst at CSE. 'Among these so-called experts on global warming are a plastic surgeon, two landscape architects, one hotel administrator, a gynecologist, seven linguists, and even one person whose academic background is in traditional Chinese medicine.'"

"Study Says 'Scientific Consensus' on Global Warming Treaty Is Just Hot Air," The Heartland Institute, December 1, 1997

is willing to employ to protect its myth of consensus—in this case, they portrayed Gore's mentor as a drooling old fool unfit to comment on the issues. Odd how such a display of ill humor is intended to humanize Gore.

University of Virginia professor emeritus S. Fred Singer details his collaboration with Revelle and one other author on a 1991 article in the first *Cosmos*[21] journal "What to do about Greenhouse Warming: Look Before You Leap."[22]

Writes Singer, "Our main conclusion was a simple message: 'The scientific base for a greenhouse warming is too uncertain to justify drastic action at this time.'" This hardly fits with Gore's fable. The article embarrassed the politically budding Gore, what with it being 1992 and he having made his bones as a bestselling, if rather fevered, environmentalist sage. Apparently acting on Gore's instructions, as documents would later indicate, his aides and associates began working on Dr. Singer, using hardball tactics beyond the norm for the academic community—a crowd known for its vicious infighting ("because the stakes are so low" as Henry Kissinger put it, failing to predict the academy's financial heyday to spring from global warming alarmism).

But this foreshadowed future tainting of science with politics.

Though Revelle died three months after the article was published, he remained active in his field until the very end. No significant attention was paid this article until Gore began spinning a moving tale of how he came to his calling of planetary salvation. Gregg Easterbrook, at the time contributing editor to *Newsweek*, referred to the article and noted the political angle in a piece in the unofficial newsletter of the Al Gore Fan Club, the *New Republic*.[23]

Most damning, Easterbrook loosed a nasty little kitten from its bag: "*Earth in the Balance* does not mention that before his death last year, Revelle published a paper that concludes, 'The scientific base for greenhouse warming is too uncertain to justify drastic action at this time. There is little risk in delaying policy responses.'" Ouch.

Columnists on the Left and the Right ran with this contradiction, seeing how it offered such potential for intrigue in the otherwise dreary harangue that is green politics (typically, one side hectors about a looming parade of horribles; the other, too intimidated over the prospect of seeming to not "care," submissively agrees). This prompted a stream of intimidation and *ad hominem* attacks that have since become the hallmark of the modern environmentalist mafia. Dr. Singer relates that, "[w]hen the difference between Senator Gore's book and Dr. Revelle's article was raised during the 1992 vice presidential debate, Senator Gore deflected it, sputtering that Dr. Revelle's views had been 'taken completely out of context.'"[24] This wasn't true.

Consensus, 1975-Style

"[T]he world's climatologists are agreed....Once the freeze starts, it will be too late."

Douglas Colligan in *Science Digest*, 1975

Demands soon issued by telephone call and in writing that Singer remove Revelle's name from the article, *post mortem*. When Singer refused—for obvious professional, legal, and ethical reasons—Team Gore launched a campaign of unsavory smear tactics. These included a written suggestion to professional colleagues that Revelle had not really been an author, and that Singer had put Revelle's name on the piece "over his objections." When going after Singer proved insufficient they went after Revelle, too. Team Gore alleged that Singer had pressured an aging and sick colleague—suggesting not just coercion on Singer's part, but that the object of Gore's (current, public) adoration was actually out of his mind when his name was, somehow, affixed as co-author on the piece counseling against climate alarmism. Then they began pressuring the article's publisher to drop the piece from further distribution.[25]

All of these sleazy efforts failed. Ultimately, Singer sued, won, and received a retraction and apology.[26] In the process Singer also got his hands on documents quite embarrassing to the Gore team exposing the ugly, yet little reported, side of Gore's Revelle discipleship, which he fully discusses in a chapter he contributed to the 2003 book *Politicizing Science*.

In sum, Al Gore has for nearly two decades offered weepy tribute to a man at whose feet he learned so much about the horrors of energy use on the planet. Except that he didn't learn there what he says at all, and Gore actually tried to finally silence the deceased Revelle.

But this is by no means the end of the story of Al Gore and intimidation of those who dare disagree that *"the debate [sic] is over; we must act now!"*

More heretics to burn

Next up on the "hit" parade for Al Gore et al. after Gore's inauguration was to chase Dr. William Happer out of the U.S. Department of Energy. Though at first asked to stay on as director of energy research by the Clinton White House, Happer subsequently made the mistake of disputing Gore. In *Reason* magazine at the time, journalist Ron Bailey told the tale of Happer's fall.[27]

Bailey focuses on Happer's appearance before a House subcommittee, in which he delivered "cautious testimony... at odds with Gore's alarmist views." Specifically, Happer uttered this scandalous sentence: "I think that there probably has been some exaggeration of the dangers of ozone and global climate change."

Possibly Happer was thinking of the part in *Earth in the Balance* where Gore writes about chlorine from Manmade refrigerants called chlorofluorocarbons (CFCs), "Like an acid, it burns a hole in the Earth's protective ozone shield."[28]

No one but Happer knows. However, following this testimony, Happer says, "I was told that science was not going to intrude on policy," and that he had made his way onto the "enemies list" of Gore aide Katie McGinty.[29]

A Book You're Not Supposed to Read

Politicizing Science: The Alchemy of Policy-making, edited by Michael Gough, Hoover Institution Press, 2003.

This sorry episode later made its own way into the record of the U.S. EPA in a formal public comment:

As is described in detail in *Physics Today*, June, 1993, page 89ff, Dr. William Happer, at the time the director of energy research at the Department of Energy was dismissed from his post after opposing the prevailing views of Al Gore and his environmental aides on the issues of ozone depletion. Happer was not your run-of-the-mill appointee, but a former physics professor at Princeton University with impressive credentials. Happer was an honest scientist in a sea of green apocalyptics who surrounded Gore. He did not share this vision and his views ran counter to many of the claims of Gore as found in his book, *Earth in the Balance.* Happer proposed a UV monitoring program to measure the ground levels of UV radiation around the U.S.

The existing data at the time did not support the hysterical stories of increased UV at the Earth's surface, the skin cancer stories, the sheep going blind, *etc.* Happer simply proposed to get more and better groundlevel UV data to resolve the scare stories. He was subsequently told that his services were no longer needed.[30]

As these instances reveal, sometimes silencing a critic is insufficient, and they must be smeared. Consider the case of former chief of staff of the White House Council on Environmental Quality, Phil Cooney.[31] Unlike Roger Revelle, Cooney is alive. Like Revelle, however, Cooney was and continues to be subjected to the shameful, tawdry treatment by Gore and the greens for which they are rather notorious. Expect Cooney to be a target of the investigation-happy Democratic Congress.

As part of his job, like his predecessors, Cooney reviewed administration publications on climate change. In the course of these duties he reconciled—or, to the hysterical media, "doctored"—the administration's publications to reflect the state of the science published by the international bodies (that the alarmists tout as irreproachable). For example, a rough draft of one document admitted some "uncertainties," and Cooney, consistent with the science, changed the phrase to "significant and fundamental uncertainties." This way of putting things was unacceptable to the ears of alarmists, in that such frankness harms their policy agenda. When Cooney came across some ruminations that global warming would melt the Arctic and hurt the native populations, Cooney also nixed it as "speculative findings and musing." *But, I said "native populations!" That's serious!* Actually, that's code for *tell an unadulterated sob story no matter what the facts.*

It is entirely safe to state that on the merits the edits for which Cooney came under attack are not only supportable but, empirically speaking, unassailable improvements. Note that I cite these two examples not because they were the lesser among Cooney's purported "dozens" of evils, but instead because they happen to be those highlighted by media outlets apoplectic over the mere fact of Cooney's involvement in what they desperately though absolutely inappropriately desire to be their sole province: official statements on federally funded climate science. Specifically, I invoke the treatment on the UK's Independent Television Network (ITN), whose package on the issue included these two examples, and only these two, with a

"'The entire global scientific community has a consensus on the question that human beings are responsible for global warming and [President Bush] has today again expressed personal doubt that is true,' [former Vice President Al] Gore said in an Associated Press interview from France where he attended the Cannes Film Festival."

USA Today, May 23, 2006

spinning planet Earth projected on the screen behind the newsreader and on the chyron below him when that screen wasn't visible. Oh, and the planet was *on fire.*

That Cooney, "a former oil industry lobbyist," dared lay hands on such sacred texts was the offense, dishonestly derided as "undermining the credibility and integrity of" a government science program, "taking out stuff . . . because it conveys a way of talking and thinking about the subject that just doesn't suit the White House politically," and "not simply editing a policy statement [but] going in and altering the conclusions of scientific analysis to mean something quite different than what they—the scientists meant to say."[32] These charges made it into White House press briefings, left-wing pulpits like Bill Moyers's *NOW* on PBS, and the front page of the *New York Times.*

When I appeared on that ITN program with the burning Earth, I noted to the host's deep shock—and promises to double-check my claims—that the problem with such hysterical assertions, as usual, is that scrutiny is

"There is some irony in the fact that Vice President Gore—one of the most scientifically literate men to sit in the White House in this century—[is] resorting to political means to achieve what should ultimately be resolved on a purely scientific basis. . . . But the issues have to be debated and settled on scientific grounds, not politics. . . . The measure of good science is neither the politics of the scientist nor the people with whom the scientist associates—it is the immersion of hypotheses into the acid of truth. That's the hard way to do it, but it's the only way that works."

Ted Koppel, ABC News, *Nightline*, February 24, 1994, bemoaning Al Gore's strongarm tactics against those who dare disagree with him

not kind to them. You see, when Cooney was "repeatedly edit[ing] government documents so as to question the link between fuel emissions and climate change,"[33] (note that neither example cited by ITN in any way relates to such a claim, but consistency matters not when smearing someone) the positions that Cooney supposedly was taking from the oil industry really came from UN's International Panel on Climate Change. Specifically, Cooney reconciled the Bush document with the IPCC's chapter 12 on "Attribution" of climate change, and *borrowed the IPCC's conclusions on "uncertainties"*! Cooney's crime, then, was digging into the scientific part of the IPCC's paper, unforgivably circumventing the "Summary for Policymakers," which was drafted by the politicians and misrepresents the underlying, actual work that the participating scientists did in fact sign off on. Cooney did not allow the IPCC politicians, pressure groups, and bureaucrats to infect U.S. documents with their "sexing up" of the issue. For that, he was roundly and viciously condemned for purportedly "sexing it down." The nerve.[34]

For the sin of bringing an official product of the U.S. government into harmony with the IPCC Third Assessment Report (a largely odious piece of work which, by the way, Gore otherwise adores and cites madly when it suits him), Cooney had to be subjected to efforts to ruin him professionally.

Gore, naturally, weighed in with typical accuracy. Amid a sea of *ad hominem* attacks on pages 264–65 of *AIT*, the book, Gore claims that Cooney was installed "by the president to edit and censor the official assessments of global warming from the EPA and other parts of the federal government . . . [and] diligently edited out any mention of the dangers global warming poses to the American people." Well, not quite.

Clearly Gore is not alone in his attack machine, however. In fact, such treatment of those who dare disagree is endemic among activist organizations and their members. See the final chapter of economist Julian Simon's *The Ultimate Resource 2*, an "Epilogue" entitled "My Critics and I" in which Simon details depressing instances of what he terms "the human propensity to suppress opposing views."[35] Simon noted that "[t]he volume of substantive negative comment...has been small compared to the volume of *ad hominem* attack," which he details and which is ugly. No single example will do, though Simon joked that he might plaster on his book jacket a particularly vitriolic comment from Lord Robert May—May responded by threatening to sue. Simon responded: "There does seem to be something funny about you wanting to sue me to prevent me from printing the ugly things you say about me." Simon despaired to his readers, while also asserting pleasure in teasing such critics with their own words, "I hope it induces you to imagine what it would do to you to have so many people respond to your work in this fashion."

How to Achieve Consensus

"Some of this noise won't stop until some of these ["skeptic"] scientists are dead."

Gore guru **James Hansen**, quoted by the Associated Press, September 24, 2006

The greens even eat their own. Consider Easterbrook, who emerged as a left-of-center eco-contrarian with his 1995 book entitled *A Moment on Earth: The Coming Age of Environmental Optimism.* That last word could simply not be tolerated, and Easterbrook became publicly reviled by environmentalists for his sins of deviating from the hymnal of doom. For example, *Grist* writer Amanda Griscom slammed Easterbrook with the wildly substantive claim that he offers "too many transparently preposterous statements to eviscerate them all here."[36] Well, I'm convinced. (While I mock this claim I also sympathize, having encountered material meeting such a description, while researching for this chapter, in fact.)

The *Wall Street Journal* wrote of the response to Easterbrook, "All this has made him a target of the environmental establishment. The Environmental Defense Fund attacked his book. The book mistakenly mentioned in passing that EDF, which is supported by many corporations, had accepted money from McDonald's for helping them change their packaging, and Mr. Easterbrook agreed to include a correction sheet with each copy. But he thinks the real reason for EDF's overreaction was that his book criticizes environmental groups for perpetuating their doomsday rhetoric to benefit fund-raising and sustain them as Beltway players."[37]

Most notorious is the case of Danish academic Bjorn Lomborg, whose saga (touched on above) is too long to rehash in detail here but which is well documented elsewhere. In short, Lomborg dared research the greens' "litany" of gloom-and-doom, only to discover its falsehood. In return, he

What's the Deal?

"There are two main camps on global warming—the true believers and the 'skeptics.' The true believers are committed to a global warming creed….On other hand, global warming skeptics may reject all, some, or only one of these beliefs….

Some of my colleagues…acknowledge that the earth is warming, but insist that such warming (and cooling) is nothing unusual, and it's not catastrophic. The end result is that the skeptics tend to be tolerant of dispute and dissent because we do not necessarily agree among ourselves. The believers are not only intolerant of dissent—they are convinced that all skeptics must be motivated by greed or other evil forces….

The 'believers' even attack skeptical groups…because we may not agree among ourselves. They see this as a weakness. They are angry because it undermines their belief that we are all paid stooges of Big Oil."

—**Owen McShane,** director, Centre for Resource Management Studies and co-founder of the New Zealand Climate Science Coalition was subjected to physical and verbal abuse and even professional sanction by a Danish "Committees [sic] on Scientific Dishonesty," claiming that he engaged in that practice and assailing even Lomborg's beliefs, including his conclusion that warming will be at the low end of computer model projections—a conclusion which also happens to acknowledge the three decades of available atmospheric observational data. The government hired four activist greens, including even a population alarmist, to make the case.[38] Ultimately, the Danish Ministry of Science, Technology and Innovation, tasked with reviewing the attack, vindicated Lomborg. Oddly, this vindication failed to capture the sensational media treatment afforded the untrue claims against him.[39]

This latter reality, alas, is yet another pattern in the sorry saga of a green movement that cannot tolerate dissent, debate, or the scientific method.

Blinded by science

To create consensus, of course, firing, smearing, and airbrushing dissent is not enough—you also sometimes need to scrape the bottom of the barrel to find *consent*. That is, after you've gotten rid of the people who disagree, go ahead and manufacture supposed experts who actually *agree*.

While one need not be a "climate scientist" to contribute to a debate on climate change, the alarmists' experts are often falsely touted as experts on climate. The experts invoked to proclaim alarm often are merely experts in that which purportedly would be *impacted* by the outcome the alarmists predict. That is, the alarmists tell us all that the planet will heat up, and then they bring out an owl specialist to say what a hot planet would do to owls. Voters get an earful of scary stories about the extinctions, migration, and so on *in the event* the alarmists are right.

No such claims by a botanist, economist, et al., make the catastrophic warming any more *likely* to come true, yet their assessments are oddly treated by the media as further evidence of global warming. Similarly, cooperative economists make alarmist claims and downplay the costs of the desired agenda to increase its appeal and perception of its necessity. They, too, are often cited as "climate scientists" (a practice that "skeptics" now cheekily mimic, in hopes that the greens will demand that practice stop).

When it suits the alarmists' needs, a geologist suffices as a "climate scientist." As does a lepidopterist (butterfly expert). In the past, greens have offered as experts a hotel administrator, landscape architect, and gynecologist. As noted, they also try to discredit petitions of real scientists by sneaking phony names on them then shrilly pointing to the surreptitiously planted "Spice Girl" or some such as proof that the whole enterprise is a sham.

The sole requirement for an honorary doctorate in climatology at Green U seems to be agreeing with the alarmist agenda. If vice presidents, economists, and bureaucrats can pose as scientists, fairness demands that actual *scientists* of all stripes be allowed to propose policies, and issue economic projections regarding global warming. The alarmists comply.

Whaddya Mean You're Not Convinced!

"Protesters Call for Resignations"

"To anyone who spent time watching hurricane forecasts last summer, Max Mayfield may seem like a hero. The director of the National Hurricane Center predicted many of the season's worst storms. But a day before the start of the 2006 hurricane season, environmental groups called for Mayfield and other officials at the National Oceanic and Atmospheric Administration, or NOAA, to resign.... Mayfield put the blame on natural climate cycles when he testified before Congress in September 2005."

The furor was apparently because Mayfield remains open to the evidence, unlike the high priests of global warming: "'I'm willing to be convinced either way here,' Mayfield told ABC's Ned Potter. 'I'm always looking forward to looking at new data. If I get convinced, so be it. But I'm not convinced yet.'"

ABC News.com, May 31, 2006. (This ritual was serially repeated outside NOAA offices.)

These stances might be fine as terms of the debate if they applied to both sides. In the alarmists' minds, however, they do not. In fact, given that no credible person can disagree with them, none of those who disagree are credible, regardless of their expertise.

Consider the case of Dr. William Gray, the originator of seasonal climate forecasts and the man who pioneered hurricane forecasting. Gray is a particular target of green ire, being a credible scientist with relevant expertise and, apparently, too great an interest in maintaining science's remaining integrity. The angry "warmers" take their vitriol to absurd depths bordering on psychosis when it comes to him. A contributor to the leading alarmist climate blog claimed he would eat his hat if Dr. Gray, he of the National Hurricane Center, were more expert on global warming than Al Gore. Gore, readers may recall, earned below-average marks in the natural sciences in pursuit of his only academic degree—a B.A. in Government received with "generally middling grades."

One alarmist blog even lards its derision of Dr. Gray as "not a hurricane scientist" with comments by one Rick Piltz calling Gray "intellectually marginal."[40] Piltz's scientific training happens to be in *political* science, apparently explaining his leadership of something called the "Climate Science Watch." However, Piltz garners his letters by saying the necessary things about Bush "censoring" science (cowed, according to Piltz, under the weight of two lawsuits filed by yours truly). It would seem fair then that some obscure climate scien-

> "**G**lobal warming science by consensus, with appeals to United Nations panels and other agencies as authorities, is the apotheosis of the century-long crusade to overthrow the foundations of modern science and replace them with collectivist social theories of science. 'Where a specific body of knowledge is recognized and accepted by a body of scientists, there would seem to be a need to regard that acceptance as a matter of contingent fact,' writes [Barry] Barnes. This means that knowledge is 'undetermined by experience.' It takes us 'away from an individualistic rationalist account of evaluation towards a collectivist conventionalist account.'"
>
> **Terence Corcoran**,
> *Financial Post*,
> June 16, 2006

tist (the real kind) should host CNN's *Inside Politics*, though Piltz does seem to already be working that corner.

Now, Al Gore may be many things, if not that which he hoped when beginning his subsequently aborted graduate education in two separate fields (law and divinity, not atmospheric science). While having clearly dedicated time and passion to the issue, Al Gore is no scientist. Is Dr. Gray a "climate scientist"? Not if that means one of the eighty Ph.D. climatologists in America (there aren't "thousands!" in America, or even the world, no matter how many times a newsreader says otherwise). Yet turn the tables momentarily. The pantheon of alarmist scientists largely also fall outside the universe of climatologists, though one would never know this from the green camp's hypocrisy over who meets their threshold to opine on climate.

Their rock star Dr. James Hansen was trained as an astronomer and chemist and his doctoral thesis was on the atmosphere of Venus, which makes him sort of a climate scientist. Stephen Schneider of cooling/warming/"tell scary stories" fame is a biology professor. The UK's version of Hansen would be Sir David "*greater threat than terrorism*" King, who is a chemist. Lots of the alarmists are computer modelers, that is, they make expensive mathematical guesses though with minimal background in the relevant sciences. They hold the proper views, however, and therefore are regularly touted as "climate science" authorities.

Now it is true that, depending on the claim made, climate science is often *not* the expertise most relevant to a given matter, as made manifest in these pages in "Skating on thin ice: The hockey stick scandal." Often the art and science of statistical analysis is paramount. As such, in those instances the alarmists generally remain true to form and make clear that no statisticians need apply (Lomborg was derided as "just a statistician").

When it comes to this matter of storm formation, experts from the discipline itself would seem more appropriate authorities than a "recovering politician." William Gray remains the leading expert, of

Crushing Dissent

"The next IPCC report should give people the final push that they need to take action and we can't have people trying to undermine it."

September 2006 statement by Royal Society of London demanding that those who disagree with them stop

whom the media could not get enough in the 2005 hurricane season (until, that is, discussion turned to blaming Man, at which Gray scoffed). Gray's personal credibility being unassailable, the American Geophysical Union reacted in 2006 to his open dissent to Gore's alarmist movie by publishing a piece claiming in essence that hurricane forecasters don't know what they're talking about when making their forecasts.[41]

Meanwhile, the same crowd demands fealty to 100-year weather forecasts by people who switch with the weather from airtight, indisputable, "*consensus!*" certitudes over cooling to certainty over warming on the basis of computer models completely subject to their inputs and proven regularly and spectacularly wrong.

The Gray example is unfortunately neither isolated in tone nor approach, and is clearly not science but politics. Regardless of such *ad hominem* campaigns as experienced by Gray, his expertise and research over decades is relevant. Al Gore's expertise lies in proclaiming environmental catastrophe and self-promotion.

What judgment about these matters must a man possess to make such a statement as he would eat his hat were Gray more qualified than Gore? Is the real qualification for expertise that one agrees with alarmists? As becomes apparent throughout this book, the more the alarmists speak, the more they weaken their case.

Chapter Six

* * * * * * *

GETTING HOT IN HERE?

Before the alarmists can convince you about melting ice caps, killer hurricanes, or gloomy Eskimos, they must convince you that currently things are historically hot, and getting hotter. Toward this end, employing actual facts is not nearly as effective a tactic as gerrymandering evidence, crafting obedient computer models through engineered assumptions, changing the past, and hanging onto discredited paradigms.

The alarmists do this quite well, and the media comply. You can't blame the editors that much. Despite implicit and occasionally explicit antipathy toward capitalism, they are in the business to sell papers and draw viewers at a profit, and "Many Causes Likely at Play in Century-Long Temperature Increase of 0.6° C" won't sell copy like, say, "Be Worried. Be Very Worried."

All signs indicate that warming in some places of the planet is outpacing cooling in other places if you measure from the 1970s or from 1900. Before going any further down this road, it is important to consider two caveats: (1) "Global temperature" is a made-up concept. All we have are averages of all our different thermometers; (2) If you set your baseline somewhere else—say 1998 or 1934—the planet appears to be in a cooling trend.

These are obnoxious points for the greens and the media, who—to borrow a phrase—"don't do nuance." They *know* we're warming, and so they try to present the information in such a way to convince us.

Guess what?

* We have just emerged from the Little Ice Age.

* The slight "global warming" forgot to include the Southern Hemisphere.

* The early twentieth-century warming was more rapid than the more modern warming, in between which was a cooling.

* The Medieval Warm Period was warmer than today.

The deception about the present temperature starts at the foundation: the thermometers.

Forgetting Siberia

Imagine if you were tasked with measuring and tracking the global average per capita income, and so you set up offices in all parts of the world. Every year, you would take the numbers for all of your thousands of offices and figure out a global average.

Then imagine if one year, hundreds of your offices including many in Africa, shut down, and so you simply got no information from these countries. Would you be surprised if you added up all your numbers that year, and suddenly your "average per capita income" was higher? Would you consider that data reliable? Would you expect some media skepticism if suddenly people read your numbers and declared that world was getting much richer?

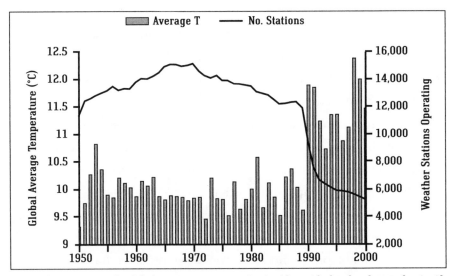

The spike in measured "global mean temperature" coincides with the shutdown of a significant portion of the world's measuring stations, many of them cold-weather stations. (Source: Ross McKitrick)

> **"T**he bad news is that the climate models on which so much effort is expended are unreliable because they still use fudge-factors rather than physics to represent important things like evaporation and convection, clouds and rainfall. Besides the general prevalence of fudge-factors, the latest and biggest climate models have other defects that make them unreliable. With one exception, they do not predict the existence of El Niño. Since El Niño is a major feature of the observed climate, any model that fails to predict it is clearly deficient. The bad news does not mean that climate models are worthless. They are, as Manabe said thirty years ago, essential tools for understanding climate. They are not yet adequate tools for predicting climate."
>
> Princeton physicist **Freeman Dyson**

When an analogous course of events unfolded in the world of climate science, the skepticism was notably absent.

From 1989 until 1992, the Soviet Union rapidly collapsed and then disappeared. While worrying about coups, orphaned nuclear weapons, more coups, and Chechen violence, they didn't do a great job of keeping up their temperature measuring stations. Thousands of Russian measuring stations closed, many of them in cold regions, as did many others around the world at the same time.

The decade that followed is now known as the "hottest decade" ever. It turns out the decade not only fails to live up to the title, but it coincided with the closing of a huge portion of surface measuring stations. Check out the graph, left, and ask yourself how not one journalist can be bothered to raise the link.

Cold below the belt

Again, the biggest problem with "catastrophic Manmade global warming" is that the warming we are seeing (derived from an average of

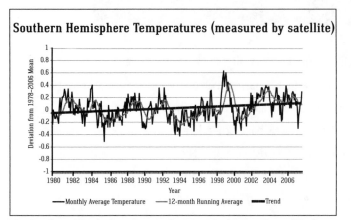

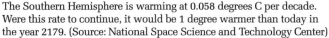

The Southern Hemisphere is warming at 0.058 degrees C per decade. Were this rate to continue, it would be 1 degree warmer than today in the year 2179. (Source: National Space Science and Technology Center)

measurements) is not demonstrably catastrophic, Manmade, or global. For example, the atmospheric temperature of the Southern Hemisphere seems to be remaining flat, on average.

Harvard University physicist Lubos Motl is a string theorist (one of the guys working on a unified theory of *everything*). On his website, he comments that the Southern Hemisphere doesn't seem to have gotten the memo, as it just isn't complying.[1]

That is, the satellite data as of September 2006 suggest virtually no warming in the Southern Hemisphere: 0.05° C per decade since the records began in the 1970s. That demonstrates that global warming isn't global. In other words, even if the planet is getting warmer *on average*, it's not getting warmer *everywhere*. In fact, the measuring station at the South Pole shows a distinctive cooling trend.

But CO_2 concentrations in the Southern Hemisphere do not deviate much at all from CO_2 concentrations in the Northern Hemisphere, according to the UN's climate change panel. In both hemispheres, the UN IPCC's 2001 report says, CO_2 concentrations had risen from about 330 parts per million to about 360 parts per million since the late 1970s. With the same rise in CO_2 concentrations, why has the Southern Hemisphere stayed flat while the Northern Hemisphere appears to have warmed?

One radical thesis is that CO_2 concentrations may not be the primary factor affecting temperature. The Southern Hemisphere is mostly ocean, and the land is less developed (paved). Paved cities are hotter than grassy fields. That factor might make explain the differences in temperature—

or at least the *measured* temperature—between the top of the planet and the bottom. If *global* warming is caused by CO_2, it seems it would be equal to the Southern Hemisphere's warming of 0.05° C per decade, while the Northern Hemisphere seems to be experiencing a *regional* warming of 0.2° C.[2]

One two-hundredth of a degree per year is, of course, a problem for the alarmist agenda pinning their supranational dreams on "global warming."

As a student of the climate debate, Motl posits that darker forces may be at work. "Normally, I would think that one should conclude that according to the observations, there is no discernible recent warming on the Southern Hemisphere, and an experimental refutation of a far-reaching hypothesis by a whole hemisphere is a good enough reason to avoid the adjective 'global' for the observed warming. Of course, the proponents of the 'global warming theory' will use a different logic. The troposphere of the Southern Hemisphere is bribed by the evil oil corporations, and even if it were not, the data from the Southern Hemisphere can't diminish the perfect consensus of all the hemispheres of our blue planet: the debate is over."

The United States of hotness?

As demonstrated with the story of the 1990s drop-off in measuring stations, we do not measure temperature everywhere—we measure where we have thermometers. This means that deploying new technologies, or new sensors, can suddenly produce (on paper) a heat wave.

The flaws in this system are obvious, unless you are a journalist or politician. Also, environmentalists have long relied upon a susceptibility to the idea that while things might seem fine here, they are simply awful just over the literal or figurative horizon. Yet they are also no strangers to trying to convince Americans every summer that the heat is unprecedented.

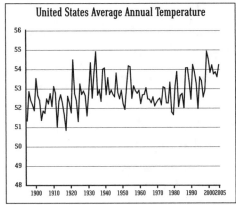

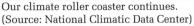

Our climate roller coaster continues.
(Source: National Climatic Data Center)

The National Oceanic and Atmospheric Administration (NOAA) maintains the database of annual U.S. temperatures from 1895 to 2005. Because, unlike Russia's government, our weather agencies don't collapse (far from it, they receive a huge chunk of the $5 *billion* annual climate lucre), and because our instrumentation is high quality, well distributed, and fairly (though not always) consistent, U.S. average temperatures are least likely to be corrupted. Temperature data here is likely the most accurate in the world. NOAA's figures[3] show:

a. Things got warmer from 1895 to about 1940. The steepest warming trend is from about 1910 to 1935. This was before significant use of fossil fuels.

b. Then the U.S. cooled off. Three and a half decades of falling temperatures spawned the "global cooling" panic, which hit its peak in the mid-1970s, just as temperatures hit their nadir. Note that this was the period of greatest growth in fossil fuel consumption (and this panic why we have satellites and radiosondes confirming the absence of catastrophic warming, today).

c. 1934 and 1998 are the warmest two years on record. 1934 was at the height of the Midwestern "Dust Bowl" and 1998 was the El Niño spike.

d. From 1975 to 1998, the country warmed. Fossil fuel consumption continued to grow in this time, but the sun also became more active (which nobody has yet attributed to earthly CO_2 emissions).

e. If, like the greens, we are willing to cite a short period of time in order to claim a long-term trend, then a possible cooling trend began in 1998, despite further, massive world-wide increases in fossil fuel use thanks in great part to growth in China and India.

f. The rate (slope) of warming from 1910 to 1934 (a period of limited fossil fuel consumption) is steeper than the rate of the warming trend from 1975 to 1998 (a period of significant fossil fuel consumption).[4]

In short, temperature is always changing, often with noticeable trends, but never with one clear cause, and not correlative with fossil fuel consumption or GHG concentrations.

Nice figures

Divide each year in four, as Mother Earth seems to do, and consider this detailed analysis of seasonal U.S. temperature variations since 1930 (the approximate beginning of large-scale fossil fuel consumption; should we wish to play the greens' baseline games, we could begin in 1935 to further illustrate summer and fall cooling and mitigate the winter and spring warming). Two seasons show a slight cooling trend (summer and fall) and two a small warming trend (winter and spring). While winter and spring are warming more than summer and fall are cooling, even the strongest warming of 0.16 degree Fahrenheit per decade is well below the bare minimum, the absolute basement of the yearly *average* that the alarmists tell us models predict. Obviously, this data does not support the theory of discernible human-induced—let alone catastrophic—warming.

Airbrushing the past

The idea that it is presently quite warm is indispensable to the Kyoto establishment's demand for urgent, expensive (and climatically meaningless)

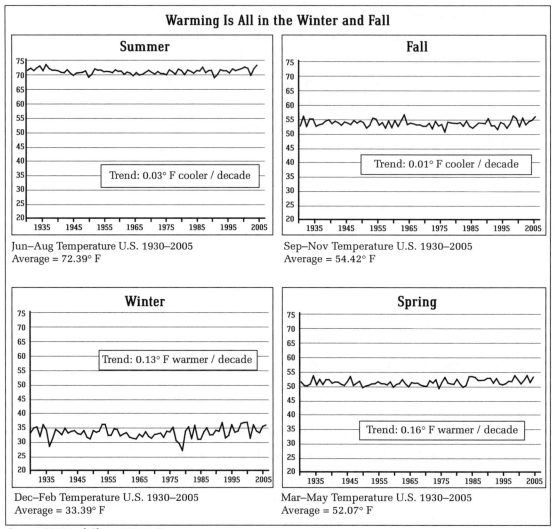

Warming Is All in the Winter and Fall

Summer

Trend: 0.03° F cooler / decade

Jun–Aug Temperature U.S. 1930–2005
Average = 72.39° F

Fall

Trend: 0.01° F cooler / decade

Sep–Nov Temperature U.S. 1930–2005
Average = 54.42° F

Winter

Trend: 0.13° F warmer / decade

Dec–Feb Temperature U.S. 1930–2005
Average = 33.39° F

Spring

Trend: 0.16° F warmer / decade

Mar–May Temperature U.S. 1930–2005
Average = 52.07° F

Source: National Climatic Data Center

government intervention on the grounds that the science is settled. In truth, the twentieth century is neither unprecedented in its warmth nor historically aberrant.

Let's start by looking back over the past millennium of the world's climate history. At right, is how it looked a mere decade back, when reconstructed in the UN's "IPCC Second Assessment Report" (1995).

Like a wart, that Medieval Warm Period just sticks out there, marring the beautiful image the alarmists have tried to paint of a perfectly stable past before the Industrial Revolution. This graph, although it appeared in a UN document (as opposed to some evil oil industry–funded propaganda) was unacceptable. The only course of action was to "correct" the past. The Medieval Warming and Little Ice Age had to go. Seriously.

David Deming, an assistant professor at the University of Oklahoma's College of Geosciences, was actually told this when alarmists mistakenly welcomed him into their club (after he published a paper they misread as supporting them): "We have to get rid of the Medieval Warm Period."[5]

In place of an airbrush, the revisionist environmentalists replaced climate history with a "hockey stick." The Hockey Stick scandal is an important tale, both as an example of green shamelessness, and to put the current trends in perspective.

Revising History

In 1995, the UN's International Panel on Climate Change (IPCC) published the following chart of temperature history, showing that climate is always changing:

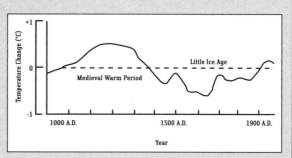

Six years later, somehow the Medieval Warm Period had conveniently disappeared, leaving the impression of a stable climate history until today. Below is the chart exactly as it appeared in the IPCC's 2001 report (at least it included error bars acknowledging some uncertainty).

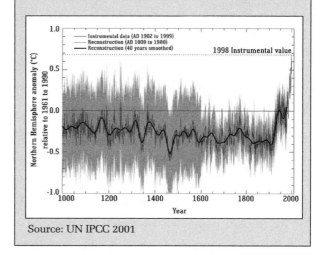

Source: UN IPCC 2001

Hockey Stick takes center ice, denies an Ice Age

A team led by Professor Michael Mann of the University of Virginia (since departed for other, dare I say "greener" pastures) published a chart in *Nature* magazine in 1998 purporting to reconstruct global temperatures, showing a stable climate for six hundred years.[6] In 1999 Mann extended the reconstruction to cover 1,000 years, showing temperature as having been stable throughout.[7] This miraculously did away with well-established climatic phenomena known as the Medieval Warm Period, followed by a Little Ice Age. These phenomena, it turned out, actually did appear in his data, but didn't find their way into his representation.

The result was the "Hockey Stick" graph—so called because it appeared to resemble a hockey stick on its side, the shaft being a 900-year straight line followed by a spike in temperature—the blade. This confirmed everything the climate alarmists hoped for. It was touted as the "smoking gun" for Manmade global warming by establishing that, until human influence, climate was largely stable.

In 2001, the "IPCC Third Assessment Report" included the Hockey Stick,[8] giving it prominent placement, in the Technical Summary, as well as the second page of the Summary for Policymakers (that section which has proven time and again to be the only one read by journalists or politicians and, as either a cause or effect of that truth, chock-full of alarmism not justified by the underlying work). In short, no one could possibly miss it. Of course, this reconstruction wildly contradicted the IPCC's own previous report, as well as extensive history and climate scholarship.

What about the well-known tales of murderous Scandinavians shipping off to a life of agriculture on Greenland during the Medieval Warm Period? What about the lithographs of children skating on frozen bodies of water throughout Europe, and Frost Fair on the frozen Thames?

These undeniable past warmings and coolings were declared regional phenomena! An easy claim, possibly, what with the absence of such images of Africans or Asians known to be in wide circulation.

So, clearly those skeptics asking such questions were despicable Euro-centrists: *just because something happened in Europe doesn't mean it happened anywhere else* (so much for climate being global). If you discard any deviations as irrelevant anomalies, you've got a straight line, and a Hockey Stick.

The Hockey Stick was now dogma. If a critic asked how this stable past accounted for, say, the Frost Fairs well established in pictorials as having taken place on an occasionally ice-bound Thames, the obscuring reply came: *see, it's warmer now*! These days, of course, when we get regional coolings, it is evidence of global warming. The Green Party USA explained in 2006, "Another probable consequence of carbon dioxide induced climate is that—due to the complex nature of the total change process—some areas of the globe will experience local cooling rather than warming."[9] It must be nice to assert a hypothesis that can be proven by any imaginable evidence.

More upsetting, the greens were parroting claims by actual scientists.[10] At minimum, regional cooling is routinely explained away as being the result of the very same culprit: industrial facilities.[11] The frozen Thames, the alarmists insist, was such a regional cooling, but not one that signified global warming. Keep up with me here, people.

Contradicting its own prior report is not a novel event for the UN's IPCC. In fact, the IPCC is a serially self-undermining effort with each product clearly establishing that the predecessor's claims of certainty (or close enough for policymaking) were wrong. Yet "news" stories ritually emerge instead along the lines of *now we have certainty*, or *this is proof of further certainty*. My favorite is the serial claim, not restricted to the IPCC of course, that *Report X or Paper Y only strengthens the consensus that existed; yes, we were sure before, but we're just more sure now. That's why it's on the front page. No other reason.* This ritual will be repeated with the "IPCC Fourth Assessment Report" expected to be released in four parts (for maximum media coverage) beginning in February 2007.

Political Science

Here's how the Clinton-Gore team further twisted the Hockey Stick in the run-up to the 2000 elections, also erasing the "error" bars of uncertainty to compound the program's inherent and fatal flaws later revealed:

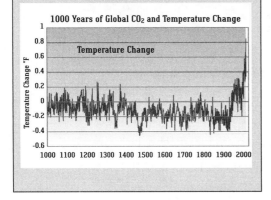

1000 Years of Global CO_2 and Temperature Change

This media treatment established one certainty without doubt—the sacred role that is played by this UN-organized body—and also accorded iconic status to the Hockey Stick itself until the latter had to itself be airbrushed out of the IPCC's pantheon of proof. The ancient Priory of Sion, dedicated to protecting the secrets of Jesus's family life as imagined by *Da Vinci Code* author Dan Brown, have nothing on the media cadre assigned to protect the secrets of the IPCC.

Although it ought to have raised eyebrows for suspiciously revising what we know about past temperatures, Mann's chart achieved iconic status at the IPCC and received similar fawning by the Kyotophiles and their media allies. Suddenly, humanity was experiencing its hottest temperatures in a millennium (perhaps a bit sheepish, the IPCC dared not actually directly say what Mann's picture clearly implied).[12] The media celebrated Mann and his chart. The debate was over. Here was the proof.

The National Academy of Sciences later debunked Mann's chart,[13] and the IPCC discarded it, but Al Gore and the mainstream media continue to hold it up as proof of a burning apocalypse.

The first problem with the Hockey Stick was that Mann's temperature readings are suspect. Remember, we did not have thermometers for the past one thousand years. Of course, scientists regularly reconstruct past temperatures from "proxy" data—items that give us a clue to past temperature, such as the size of tree rings,[14] bore hole samples, peat bogs, extracted ice cores, bristle cones, and a host of others.

For the pre-thermometer data, Mann and his "Hockey Team" used what they called a "multiproxy technique," combining many of these measures, and weighting them differently (for example, giving tree rings the most prominence). For the period when instruments were available to measure the temperature, the instrument measurements were used.

Mann's chart has a temperature spike right at the time that proxy data gives way to instrument data—that is, at the point he stopped using tree rings to tell the temperature, and started using thermometers.

Maybe this is pure coincidence. Or maybe Mann wasn't reading the tree rings correctly. Unluckily for Mann, we can check on this. If we use his multiproxy technique to measure the temperature during a period in which we *do have* instrument readings, we find that his technique does not correctly measure the temperature (unless Mann wants to argue that the thermometer readings were just all written down wrong).

As such, despite widespread use, tree ring data's utility may be limited exclusively just to comparison with other tree ring data. If Mann's tree rings don't show that things got warmer in the medieval period, it probably indicates the rings are not as sensitive as the instruments on which Mann relies for his twentieth-century temperatures. Conversely, if you use Mann's technique to measure the twentieth century, you don't find the warming our thermometers find. This alone is deeply suspect.

A second outstanding feature is the error bars. Admitting a margin of error, the gray shaded area shows where the temperature might have actually been according to the proxy data, that is, what might be a reliable range indicated by the proxies. Such humility is required for credible science, but it makes bad politics when a tremendous agenda is at stake. Accordingly, the error bars were airbrushed out of other, even more politicized treatments of the Hockey Stick, including climate science's effort to get Al Gore elected president in 2000: the *U.S. National Assessment on Climate Change*, released in November of that election year.[15] While

Mann's Hockey Team didn't do this airbrushing, they should have screamed bloody murder about it instead of acquiescing.

Third of course are the missing Little Ice Age and Medieval Climate Optimum ("warming," known as the "Optimum" for coinciding with unprecedented prosperity, well-being, and population growth from reduced mortality).

Steven McIntyre and Ross McKitrick, the Canadian mavericks, later demonstrated the complete uselessness (or supreme usefulness, in the alarmist view) of Mann's models. If you were to plug in a completely random set of data into Mann's models, you would yield the same Hockey Stick result.[16] Mann's model, then, is something like a goat: all sorts of things could go in the front end, but the same stuff will always come out the back end.

Hockey Stick in the penalty box

Veteran mining industry mathematician Steven McIntyre and economics professor Ross McKitrick (together, "M&M"), noticed obvious defects in the Hockey Stick chart and the claims it was being used to support. They asked for Mann's data to replicate the team's work, as any study or experiment, if it is "sound science," must be capable of replication. M&M concluded, on the basis of information obtained by an associate of Mann's, that the data "for the estimation of temperatures from 1400 to 1980 contains collation errors, unjustifiable truncation or extrapolation of source data, obsolete data, geographical location errors, incorrect calculation of principal components and other quality control defects."[17] Well, that would be a problem...

It used to be that the scientific method meant "proposing bold hypotheses, and exposing them to the severest criticism, in order to detect where we have erred."[18] When Sir Karl Popper described science thus three decades back, it was before such vast sums were at stake as the

U.S. government's $5 billion annual climate research budget. For their criticism, M&M felt the wrath of the global warming establishment.[19] They encountered obstacle after obstacle in their efforts to check the original data. Undeterred, after three years they had sufficiently exposed the Hockey Stick to defrock it as follows:

> *Nature* [the science journal that published the original Hockey Stick research] never verified that data were correctly listed: as it happens they weren't. *Nature* never verified that data archiving rules were followed: they weren't. *Nature* never verified that methods were accurately stated: they weren't. *Nature* never verified that stated methods yield the stated results: they don't. *Nature* undertook only minimal corrections to its publication record after notification of these things, and even allowed authors to falsely claim that their omissions on these things didn't affect their published results. The IPCC's use of the Hockey Stick was not incidental: it is prominent throughout the 2001 report. Yet they did not subject it to any independent checking.[20]

Ouch. It seems we have an establishment promoting something that simply appeared too good to be true for the future funding prospects of their discipline. This was only part right: while funding has only escalated, the Hockey Stick has certainly been proved not true. Using the original source data to correct these errors, M&M concluded that "[t]he particular 'Hockey Stick' shape . . . is primarily an artefact of poor data handling, obsolete data, and incorrect calculation of principal components."[21] The abstract of McKitrick's paper cited, below, explains the bigger-picture meaning of the controversy:

> The Hockey Stick debate is about two things. At a technical level it concerns a well-known study that characterized the state

of the Earth's climate over the past thousand years and seemed to prove a recent and unprecedented global warming.... [T]he conclusions are unsupported by the data....[Second, the] Hockey Stick story reveals that the IPCC allowed a deeply flawed study to dominate the Third Assessment Report, which suggests the possibility of bias in the Report-writing process.... [22]

In response to M&M's initial inquiries and revelations, Mann claimed that M&M used the wrong data set as well as only 112 proxies when 159 were needed. However, M&M revealed that Mann's original research paper only *contained* 112 proxies, which were the proxies Mann instructed his associate to provide to them. So, Mann assailed M&M for following his lead. He also claimed that many other paleoclimatologists have been able to replicate his results closely. Of course! Applying the same flawed methodology to the same flawed data will generally result in the same flawed results. Also, other scientists have failed to replicate Mann's results and, as noted, have used his flawed methodology to replicate his results, but with *different* (random) data. Not good.

The Hockey Stick was elevated to iconic status without scrutiny, apparently because it was just too dreamy a result to check. Possibly as a result of exposure of alarming problems such as a "Hockey Stick" shape resulting from just about any set of input data, indicating a bias in Mann's design. Mann and his colleagues in the Hockey Team continued to be unhelpful and more attentive to *ad hominem* attacks than responding to requests for data.

This dodginess was sufficient that a full-blown controversy ensued, thanks only to M&M's persistence and willingness to subject themselves to the Kyoto establishment's nastiness.

M&M's reconstruction[23] reveals in several ways how the Hockey Stick is fatally flawed. First, it shows the Medieval Warming and Little Ice Age actually appeared in Mann's *data*, though not his representation: "The Mann multiproxy data, when correctly handled, shows the twentieth century climate to be unexceptional compared to earlier centuries. This result is fully in line with the borehole evidence."[24]

The controversy divided the paleoclimate community. Some, like prominent German scientist Hans von Storch, recognized the issues raised by the determined Canadians and called for a rethink (von Storch called the Hockey Stick "trash"[25]). Others, mostly Mann's collaborators in the Hockey Team, refused to budge. A third group found greater historic variability in temperature than Mann, but used similar methodology and so found themselves caught in limbo. One of them, Jan Esper, noted in a scientific publication in late 2005 exactly what the stakes were:

> Scientist Peter Doran confirmed in 2002 that the Antarctic has been cooling while climate models all predict a strong warming. Sadly, the now-typical response to this conundrum is the authors, like Doran himself, trying to explain away their own results in op-ed pages of liberal newspapers to conform with green dogma. Being labeled a "skeptic" because your research harms the alarmist cause means fewer dinner party invites, it seems.

> [E]nhanced variability during pre-industrial times, would result in a redistribution of weight towards the role of natural factors in forcing temperature changes, thereby relatively devaluing the impact of anthropogenic emissions and affecting future predicted scenarios. If that turns out to be the case, agreements such as the Kyoto protocol that intend to reduce emissions of anthropogenic greenhouse gases, would be less effective than thought.[26]

In other words, if Mann is wrong, the alarmists lose Kyoto. Though it shouldn't be, this proved a tough call for science.

The following chart represents the corrected reconstruction of Mann et al.'s own data by M&M.

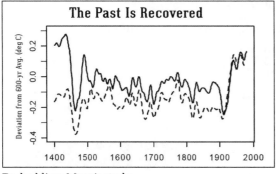

The Past Is Recovered

Dashed line: Mann's study
Solid line: corrected study
(Source: McIntyre & McKitrick)

It was now on the table that Mann's graph was a deeply flawed, but critical, tool for driving green policy that also happened to have been paid for with government money. Congressman Joe Barton inserted himself into the process demanding some transparency.[27] Barton wrote the researchers, noting that "sharing data and research results is a basic tenet of open scientific inquiry" and asked Mann for the computer code used to generate the Hockey Stick graph which had been denied to other researchers seeking to replicate the graph.[28]

Though Barton was basically asking Mann to "show his work," he incurred shrill charges of seeking to influence science through witch-hunting and McCarthyism. This was rich, coming from the Eco-McCarthyites. An association of scientists fumed that if the UN had accepted something, that ought to be good enough for the U.S. In other words, the establishment exhibited classic symptoms of having a nerve struck.

A particularly paranoid allegation came from the Republican congressman representing the Sierra Club, chairman of the House Science Committee Sherwood Boehlert. Boehlert menacingly wrote Barton, "The only conceivable explanation for the investigation is to attempt to intimidate a prominent scientist and to have Congress put its thumbs on the scales of a scientific debate....The precedent your investigation sets is truly chilling."[29] *Brrrr*!

Actually, to anyone having followed the Hockey Stick's tortured history and Mann's evasions, a rather obvious objective was to discover the validity, or lack thereof, of this emblem of alarmism. No wonder Boehlert et al., were so concerned. Ultimately, Boehlert "asked the NAS to evaluate criticisms of Mann's work, and to assess the larger issue of historical climate data reconstructions. The NAS agreed to the science committee's request, but only under terms that precluded a direct investigation of the issues that prompted the original dispute—whether Mann et al. had hidden adverse results and whether the data and methodological information necessary for replication were available."[30]

Occasional collaborators of the Hockey Stick's authors were even placed on the panel (not a complete surprise, actually).[31] By this request Boehlert obviously expected a report consistent with the recent template of NAS reports bearing uncomfortably sober conclusions that somehow evade the media's coverage, accompanied by an alarmist money quote prominently dropped into the summary—now proven to be as far as a journalist gets, if he actually makes it past the green pressure group's press release.

In June 2006 the report was issued.[32] The Academy deconstructed the Hockey Stick in the gentlest way possible given the stakes and the devastation of their actual conclusions. They strapped on the velvet glove to sorrowfully and subtly but unavoidably humiliate Mann and the Hockey Team, concluding that while *they* couldn't establish it from Mann's work, the claims of the 1990s being the warmest decade and 1998 the warmest year in the past 1000 were "*plausible.*" That is to say, *the answer's not in here, but it may be out there. Somewhere.* Ouch. That's rough stuff in the *hey-this-is-our-gravy-train-too* world of alarmist wagon-circling.

The panel reaffirmed the Little Ice Age which Mann swept from history (along with the preceding Medieval Warm Period), and made it clear that there is nothing close to the certainty Mann claims about past temperatures, nor, therefore, about the present.

That is, due to the even larger uncertainties in the proxy data prior to 1600 A.D. the panel concluded that it is fair to assume it is as warm now as at any time *in the past 400 years—that is, things have warmed up since the Little Ice Age ended.* Well, there's a groundbreaking statement! Now, with the Hockey Stick's Thousand-Year Spike being reduced by 60 percent, surely the media would at least avoid saying stupid things like this proved it was correct, with "certainty" no less, for even longer than even Mann et al. tried to pull off? Guess again. CNN actually spun this 180 degrees, asserting the NAS panel not only endorsed the "Hockey Stick" when it didn't, but for the past *2,000 years* of temperature!

Four hundred years ago, as the panel reminded us and again in repudiation of the Hockey Stick, we were in the depths of the Little Ice Age. A headline announcing that things are warmer now than they were during the Little Ice Age should therefore rival *"Sun Rises in East"* for banality. Or, as put by chairman of the Senate Environment Committee James Inhofe of Oklahoma, this truism is akin to claiming that August being hotter than January is proof of an alarming warming trend. Senator Inhofe, however, never had the benefit of journalism school.

The media treated this smackdown as an endorsement. NAS concluded that Mann's theory was not supported in the work purportedly affirming it, and merely "plausible." The report's context made clear that this was a very diplomatic but damning slur. Translated by an alarmist media "plausible," followed by a damning analysis, was morphed into "likely correct" on CNN[33] and to *"most likely correct"* in the *Boston Globe*.[34]

Days later, the Associated Press set a new standard for political advocacy couched as scientific reporting, invoking the NAS panel's slam of the Hockey Stick when reporting Senator John Kerry's "energy plan." Gushing support for the Kerry press release's obeisance to warming alarmism, AP spun wildly in stating that "a panel of scientists said last week global warming is now an *undeniable* scientific phenomenon."[35] That conclusion must have been in the Annex issued only to the press.

The good news is that Al Gore's internet, of all things, makes life just a little more difficult for these sophists, enabling us to look more closely at this case study of climate-science-in-action. Lubos Motl took to his blog to do Mann in with a dissection of the audio of the NAS panel inquiry, sparing the world the drudgery of virtual attendance at one of the most oppressively turgid fora known to Mann, er, man:

Mann suddenly started to say that he never said that he was certain that the current era is the warmest era in the last 1000 years and, on the contrary, he always emphasized that their research was meant to show how *uncertain* these numbers are. Well, we probably live in different Universes because in this Universe, he said it roughly 350 times and 870,000 articles have been written about this extraordinary statement.

Motl then provided the link to the audio files and the exact time of these false statements, writing: "You're exactly one click from verifying that various media and the RealClimate group blog are just trying to fool you completely."

The sum total of the Hockey Stick experience suggests nothing short of intent to deceive on the part of the alarmists, as well as among their colleagues in the media. The NAS panel inescapably indicted the Hockey Stick, the UN IPCC, and the Mann team itself. The NAS panel specifically repudiated three-fourths of Mann's record, specifically accused the IPCC of misrepresentation, and specifically accused the Mann team of downplaying historical uncertainties.

As if further proof were needed of media irresponsibility on these matters, with over a week to read the report and get the facts straight, even the presumably more substantive *Nature* magazine ensured its place in the campaign to perpetuate the industry's gravy train, headlining its foray "Academy affirms hockey-stick graph."[36]

The Hockey Stick is dead. Long live the Hockey Stick.

Are models really dumb?

"General circulation models," also known as GCMs or simply climate models, serve as the basis for lurid climate alarmism, for claims of future temperature, ice melt, sea levels; it's a sure thing that whatever ogre under the bridge is proffered, it is the output from a computer model. Models are hypotheses about climate behavior which, like all models, produce results that are a direct function of the assumptions plugged in and factors considered.

The left-wing online magazine Spiked.com admirably remains quite reasonable on climate alarmism, being possessed of ideology sincere enough such that they detest the use of fear to infantilize the working class.

Times Fudges a Baked Alaska

On July 1, 2002, George Mason University's VitalSTATS caught the *New York Times* in some sloppy reporting on Alaskan warming:

In Alaska, "the average temperature has risen about seven degrees over the last 30 years," according to the June 16 *New York Times*. In discussing the severe effects of climate change in Alaska, the newspaper of record observed that such "rising temperatures . . . are not a topic of debate or distraction. Mean temperatures have risen by 5° F in summer and 10° F in winter since the 1970's, federal officials say." Fortunately for Alaskans, the skyrocketing mercury was brought down to Earth by official data.

The *Times* relied on oral testimony about the effects of climate change in Alaska. And though it mentioned the recent Climate Action Report from the Environmental Protection Agency in passing, the *Times* did not appear to have consulted the actual text of the EPA report to check its alarming temperature readings (nor, it seems, had the unnamed "federal officials" who served as their sources). The report states that "warming in interior Alaska was as much as 1.6° C (about 3° F)" over the last one hundred years.

The *Times*'s assertions baffled professor Gerd Wendler and his staff at the Alaska Climate Research Center. In response, Wendler posted to the internet a data analysis of mean annual temperatures at four widely dispersed weather stations in Alaska from 1971 to 2000. The mean temperature increase for Anchorage was 2.26° F and for Nome 2.28° F.

Spiked's Rob Lyons concludes that models are not entirely without value to the debate, but exhibit many problems that include conscious and unconscious bias, incomplete data, and a very limited history. Of course, as exposed below, models in fact are without utility in their current state as even climate alarmists have accidentally proved.

Nevertheless, Lyons provides a fair and balanced explanation of climate model benefits, problems, and their relationship with other measuring systems:

> ## How to Achieve Consensus II
>
> "We have 25 or so years invested in the work. Why should I make the data available to you, when your aim is to try and find something wrong with it?"
>
> Hockey Stick co-creator **Phil Jones**, replying to an inquiring Warwick Hughes

> The physics of individual climate elements is not fully understood, particularly in relation to clouds; we don't know how much cloud will be produced in a warming world and what the net effect of that cloud will be. In addition, new announcements from research teams are made regularly about factors that hadn't been fully appreciated before.
>
> Also, models are, by their very nature, simplifications of the real world. Consider a non-climate example: the Millennium Bridge in London. This was a relatively simple system to model. But when the bridge opened in June 2000 it had to be quickly closed again because the effect of people actually walking on it caused the whole thing to "wobble." So even engineers with far less complex problems than world climate to solve can get things badly wrong....[F]or reasons quite unrelated to climate science, each new set of results and each new report is leapt upon by one side or the other as confirmation of their own position.[37]

Lyons hints at, but then fails to follow up, the dirtiest secret of all regarding climate models: when we attempt to test them, they fail miserably.

This does not seem to trouble the climate community: *they're all we have . . . oh, and they're* very *useful for producing lurid scenarios.*

Let us not quibble over this here. Instead, consider the consequence of the models' ignorance of the role played by clouds and the sun and the influence of oceans and topography.

When the Clinton-Gore administration produced its election-year "National Assessment on Climate Change"[38] in 2000 it selected two computer models to make the case. Unsurprisingly, from among the more than two dozen computer models available, the NACC used the pair that produced the hottest and wettest results.

Scandalous behavior was rife.[39] Consider that home-grown American climate models are more conservative than others in the temperature increases they project. Naturally, therefore, "[t]he [National Assessment team] rejected those models and instead selected Canadian and British models that consistently yield higher temperatures, more extreme weather events, and worse environmental disasters than any of their counterparts. One model used by the GCRP projects precipitation in the Colorado River Basin will increase by 150 percent over the next century; the other says there will be only a 5 percent increase. One predicts an 80 percent increase in precipitation for the Red River Valley; the other an 80 percent *decrease*. One even projects that most of the world's tropical forests will disappear within 50 years."[40]

One of the groups whose model was used, from the UK's Hadley Centre, had the honesty at the time to admit on its website, "In areas where coasts and mountains have significant effect on weather, scenarios based on global models will fail to capture the regional detail needed for vulnerability assessments at a national level." If lakes, seas, oceans, or mountains affect the weather where you are, which Hadley admitted means most of the world, this model won't tell you much. In other words, their model was not useful for the purpose to which it was being put.

In what ought to have ended the debate over using such predictions in policymaking or even for any taxpayer-funded purpose, the alarmists also admitted that, when asked to test themselves by looking backward and trying to reproduce past climate (which unlike the future we do know), the models performed more poorly than a table of random numbers. Climatologist Patrick Michaels, in his "Review of the 2001 U.S. Climate Action Report," addressing a Bush administration document that incorporated U.S. National Assessment findings, notes these problems in his opening paragraph: "Whatever originates from the USNA is highly flawed because the USNA is based upon a true miscarriage of science: it is based upon two models for future projections of climate that perform worse than a table of random numbers when applied to recent climate. The producers of the USNA, mainly the U.S. Global Change Research Program, have ignored this glaring problem, even as it is well-known that they were aware of it. Further, the USNA is based upon a selection of the two most extreme climate models for U.S. temperature and precipitation, for which there is no scientific defense."[41] (Those pesky random numbers foiling things again; just maybe there's nothing random about this flaw in the alarmists' high-profile weapons.)

As he testified before Congress, Michaels had asked the government to run this test, which he had conducted independently as a standard "Monte Carlo" analysis for testing the validity of a hypothesis.[42] The government replicated Michaels's results, and admitted as much, to which he alludes above. Then, as if no such knowledge had been obtained, the government modelers continued to merrily employ these useless products to project climate for purposes of informing governmental policy!

In short, models cannot "hindcast" past climate. As such, they cannot reliably forecast, the precise use to which they are put. No GCM has yet replicated the medieval or Roman climate events. The models simply are not real-world. They have been disproven as tools of climate policy. Says

Michaels in his "Review": "It is scientific malpractice to use them. I choose my words carefully here. If a physician prescribed medication that demonstrably did not work, he would lose his license."[43]

All of which must explain why each new simulation of doom and gloom makes headlines. The *New York Times* might as well give a front-page splash to a high score on Donkey Kong for all of its news value. (Actually, Donkey Kong high scores have been going up over the same time as temperatures. Hmmm. Someone needs to model that correlation.)

Although scientists and computer modelers are loath to admit it, given the enormous sums of taxpayer money given them to produce "projections"—which without protest they allow the media to portray as predictions—nobody really knows enough about long-term climate to make a model that can provide credible projections even at the continental level (in short, climate reality is too complex for any climate model to replicate). Predicting future climate on the scale of a city or state would be far more difficult,[44] but still alarmists and their policymaker allies bandy about very detailed, city-specific horror stories in order to advance an agenda.[45] Again and as betrayed by the Hadley example, modelers admit this weakness. One article in *Nature* ran under the headline, "Climate Models Have No Answer to Burning Questions" lamenting the unreliability of models and relaying scientists' calls for more taxpayer money so that they could improve their models.[46]

When we examine the UN IPCC's various models, we note that, interestingly (though not to the media, apparently), the *lowest* model projection of future warming has coincided very well with the actual temperatures over the past three decades: 0.17 degrees per decade or about *a degree and a half in the century*. Yawn. Earth's been there, done that.

When Jim Hansen testified to Congress in 1988 and started the whole tidal wave of alarmism, he presented three scenarios. Setting the tone, the media, environmentalists, and politicians focused on the one projected outcome that showed the most warming. In fact, temperatures have proven remarkably similar to the *least* dramatic of Hansen's three scenarios; that is, the temperatures are as he suggested would occur only with drastic emission reductions, despite the fact that *emissions* have proceeded consistently and pretty well on the track of what he declared as catastrophic in that it would yield the most delirious climate. No emission reductions, no catastrophe: in other words, his scenarios have been proven wrong, which, naturally, has led to cries of "Hansen was right." Hansen himself seems to believe his own press—of which he certainly gets a lot—as the idea that maybe he was wrong is not a prominent topic in his writing.

Still, the highest actual computer model projection is the one adored and propagated by the media. Even those who have had the courtesy to report "*up to* 10 degrees"[47] warming generally still mention just the outlying, most lurid, and therefore least likely scenario as that requiring reportage.

That so many models produce such a broad array of outcomes further testifies to their lack of utility. Depending upon how rapidly a future warming came about, were it to occur and for whatever reason, 1.5° C is hardly the same scenario as 10° C. Man has adapted to temperature swings throughout time. Here, a degree and a half is predicted to come in one hundred years. Consider that quick or not so quick, as you wish. But do not consider it unprecedented or catastrophic.

Models also share in common a gross overestimation of growth in global population and GDP (discussed later), far beyond anything history suggests is even remotely feasible all in order to grossly overestimate Man's contribution of GHGs and develop wild global warming scenarios. They generally assume that GHG emissions will increase—and, have increased—at twice the rate as that observed over past decades, a rate which shows no signs of increasing to meet the modelers' fantastic assumptions.

What about Al Gore's model? Clearly, he selected an outlier model projecting a temperature spike that would be large even in the fairly alarmist world of the UN. Gore reached outside the modeling mainstream to an even less credible, isolated program to claim that temperatures may increase by up to 11° C. While the computer program does present an unrealistically hot outcome, Gore severely misrepresents even the program's actual "finding." The actual paper upon which he bases his alarmism[48] suggests that the *lower* end of the projected temperature increase, the 2° C to 4° C range, is the most likely. Even this temperature increase, however, is hotter than what one can reasonably derive from the combination of models and observations about the current warming trend. That is, he is completely alone on these claims, while insisting that the universe of the informed is with him. Shameful.

By opting to present, rather *misrepresent*, the most extreme result of a rogue model, Al Gore not only abandons adherence to his beloved "consensus," but he acknowledges that *even the UN's most lurid scenario is insufficiently alarming* for his purposes of seeking his desired policy changes. Al Gore to UN: keep up!

The North Korean tiger

On top of bad science, bad statistics, and bad scruples, the alarmists employ bad economics.

In order to predict future global warming, the UN's IPCC assumes that many countries will experience economic expansion. Acknowledging that energy use is

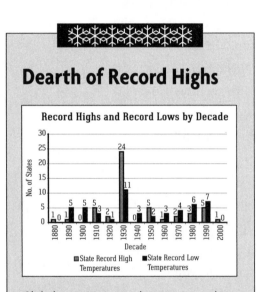

Dearth of Record Highs

Record Highs and Record Lows by Decade

■ State Record High Temperatures
■ State Record Low Temperatures

Global warming is not showing up on the hottest days. While winters and nights are getting warmer on average in the Northern Hemisphere, new record-high temperatures are not being set. In fact, South Dakota is the only U.S. state to experience its record-high temperature since 1996. Statistically speaking, in a stable climate, you would expect about three new record highs to have been set in the past decade.

Source: National Climatic Data Center

inextricable from economic growth, the IPCC then estimates that these countries will use more energy than they are using now—which means they will give off more greenhouse gases. Up to this point, it's all fine and good.

But when it comes to poor countries, the UN projects significantly greater economic development than is reasonable. This means they over-estimate the GHGs these poor countries will produce in the future. Plug bad economics into inadequate models and you get climate chaos. This means that even when accepting all of their assumptions, the alarmists must still build one fantasy on top of the entire house of cards to produce their scary tales.

Stephen Hayward, a scholar at the American Enterprise Institute, spelled out the folly of the UN's way of measuring and projecting economic growth:

> The resulting projections are frankly absurd. Playing the IPCC's own number game, by the end of this century, the per capita income of South Africans will be four times higher than that of Americans. North Koreans, Libyans, Algerians and Argentines will also have higher real per capita incomes than the United States. Except for 'proving' global warming, the IPCC method just doesn't work. It's an obvious distortion.[49]

Chapter Seven

❋❋❋❋❋❋

MELTING ICE CAPS, ANGRIER HURRICANES, AND OTHER LIES ABOUT THE WEATHER

1f you're going to give up your freedoms, your conveniences, and your affordable energy to them, they need to scare you. Every bad thing that's already happening becomes the fault of Manmade global warming. Hurricane Katrina: Global Warming. Droughts: Global Warming. Flooding: Global Warming. Too many insects: Global Warming. Too few insects: Global Warming.

The weather is now your fault.

After assigning blame, it is necessary to promise you, like the Book of Revelation, that things will get much, much worse. Hurricanes will get stronger and more frequent. Everything that hasn't already melted will. The tides will rise, drowning our cities.

Where there is a heart-wrenching tale, it is told and blamed on Man, however speciously; where no such tale exists, it is fabricated.

Can't bear the truth

A poster child of this phenomenon is the cuddly (from afar) polar bear, to which alarmists turn to warn of the horrors of a possibly warmer world. Polar bears, like penguins, are the unwitting mascots for green lobbying groups. *Time* magazine chose this baby seal–eating mammal as the cover boy for its issue declaring: "Be Worried. Be Very Worried," and

Guess what?

❋ The South Pole is getting colder.

❋ Most polar bear populations are thriving, (even if Al Gore falsely says they cannot swim).

❋ Not a single hurricane hit the U.S. in 2006.

❋ Most experts do not attribute the recent hurricane activity to greenhouse gases.

Al Gore offered a computer-generated bear flailing about for icy salvation in his movie. Claims of the imperiled polar bear run the gamut, from drowning in water to which they are unaccustomed (not true, they encounter it every summer), to starvation-induced cannibalism due to their purported inability to traverse disappearing ice to access their traditional diet.

It is certainly touching to hear environmentalists complain that not enough ringed seals are being devoured. (And it's not only bears producing disappointing kill tallies: greens actually presented a slide show at the Buenos Aires and Montreal Kyoto negotiations the complaints of which seemingly included that Inuits, too, aren't able to mortally club enough seals.) To merely call these claims overblown, however, is an insult to overblown claims everywhere.

The Associated Press disseminated a story claiming that polar bears may be "turning to cannibalism because longer seasons without ice keep them from getting to their natural food." The mid-2006 claim was based on three purported incidents of "cannibalism," all from 2004. Yet actual research reveals that the bears are thriving in those areas where there is warming, and suffering where there is cooling.[1] (Yes, Arctic cooling.) According to leading Canadian polar bear biologist Dr. Mitchell Taylor, Department of the Environment, Government of Nunavut, "Of the thirteen populations of polar bears in Canada, eleven are stable or increasing in number. They are not going extinct, or even appear to be affected at present."[2]

What about those penguins down on the other end of the planet? Despite hysterical coverage by *National Geographic*, the *New York Times* and (to a slightly more balanced extent) the BBC, it seems that their reluctance to procreate might have more to do with their modesty and fragile psyches than temperature change. Research revealed that breeding pair populations declined sharply despite no evidence of warming or cooling, but coinciding with the advent of ecotourism. You try getting it on while huge helicopters descend only to spew forth hordes of earnest, Gortexclad creatures towering over you.[3] Imagine, say, honeymooning on the tarmac of an air show in Germany.

Chilly reception for Arctic claims

Global warming models uniformly predict that the planet's overall warming will be amplified at the poles—they call this "polar amplification." These predictions apparently feed the hopeful stories of miserable wet polar bears, but they have already been proven unreliable.

As we shall see, these model predictions remain committed to the rhetoric of the 1970s climate alarmists. Yet since those days, claims of polar amplification has tempered from "greatly enhanced" to "difficult to measure" down to the present "insignificant."[4] The purported cause of polar amplification has also varied widely, from clouds and how much warmth was absorbed by other particles in the air, to how much solar heat was retained or deflected by snow and ice (technical explanations all, including "snow-ice-albedo" [feedback], to "dynamical circulation feedback" to cloud cover to soot to aerosols to tree lines), all just in the

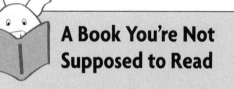

A Book You're Not Supposed to Read

Meltdown: The Predictable Distortion of Global Warming by Politicians, Science and the Media, Patrick J. Michaels, Washington, D.C., Cato Institute, 2004.

past several years. Again, it appears that climate science tends to begin with its conclusion—in this case, that polar bears will drown—and comes up with the explanation later.

This rhetorical change about polar amplification seems to reflect the ravages of mounting data eroding the theory, given that the poles are not significantly warming. The ice cover at the North Pole *is* significantly thinner than it was decades ago, and this trend seems to be continuing. But the hysterical—and provably false—cries of unprecedented open Arctic water and imminent polar meltdown go too far and go unchecked.

One key alarmist tactic is to redefine the word "Arctic." The actual Arctic Circle is at about 66 degrees and 33 minutes North Latitude, representing the tilt of the Earth, and the southernmost point to be exposed to sunlight twenty-four hours at the summer solstice (and twenty-four hours of night at the winter solstice). Some scientists define "Arctic" based on climate (and because climate is always changing, their definition of Arctic is always changing). Others delineate elsewhere.

The *Arctic Climate Impact Assessment* (ACIA),[5] which has served as the basis for serial, breathless stories about a melting Arctic in recent years, chose to expand the Arctic Circle 450 miles in all directions, setting the southern limit at 60 degrees North Latitude (pity the poor residents of the Shetland Islands who suddenly became Arctic dwellers. *I'm an Eskimo who didn't know!*) The ACIA expanded the Arctic by about 50 percent, adding about 3.9 million square miles (the equivalent in surface areas of adding the entire U.S. plus two Frances). The Arctic as they define it is two-thirds covered with ocean, but they used land-borne measuring stations to determine the temperature.

This odd definition of the Arctic—Arctic gerrymandering—either intentionally or by good fortune suited the ACIA's particular alarmism. Oregon State climatologist and researcher Dr. George Taylor, CCM (certified consulting meteorologist) specifically notes that the ACIA concluded the Arctic is melting by this redefinition of the relevant area, but with the added benefit of including several very unreliable Siberian measuring stations (some of those *not* closed).

In this context, Taylor notes the importance of the fact that parts of the Arctic have warmed while parts have cooled (again, however, the warming areas have been more hospitable to bear populations, as evidenced by their numbers).

No doubt purely by chance, the ACIA team selected 1966 as their baseline—*the absolute low point for measured temperatures* in the past hundred years—thereby exaggerating the warming of 0.38° C per decade to *four times* that experienced over the entire century. This is akin to declaring a warming trend in any given year, beginning one's measurements in January. Why, yes, it *did* subsequently warm after the coldest period! Of course, it is also consistent with the general thesis of catastrophic Manmade warming theory, that the Little Ice Age, to the extent it cannot be airbrushed from history, is a convenient starting point from which to declare a warming trend, *by definition*. What the ACIA also chose to not emphasize was that this warming was still less aggressive by 50 percent than the warming of 1918–1938, and the same as the warming experienced from 1880–1938. As such, the ACIA exercise is no more than too thinly disguised alarmist advocacy.

These Arctic temperature charts disprove the alarmist claim that the Arctic was stable until Man came along and hotted things up. They show that GHG concentrations do *not* dictate Arctic temperatures any more than they dictate

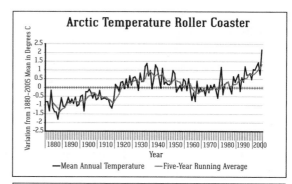

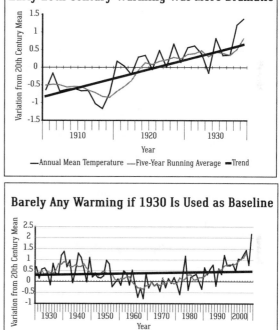

Source: Goddard Institute for Space Studies

global mean temperatures. Of course, the charts do reaffirm the importance of the baseline selected. In fact, the Arctic appears to be undergoing a long-term warming trend of a few degrees—but clearly not as an effect of Manmade GHGs in the atmosphere. If Manmade GHG increases were the cause of this long-term warming, the Arctic would not have seen a cooling from the 1930s to the 1970s.

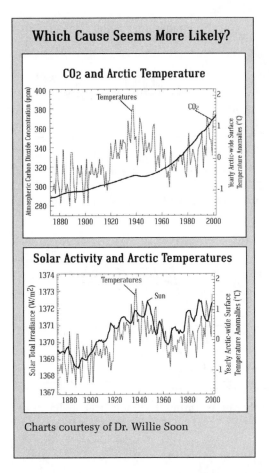

Charts courtesy of Dr. Willie Soon

What about any Arctic-specific data regarding the correlation between greenhouse gases and Arctic surface temperatures? Fortunately, we do have this information, and it doesn't look good for the alarmists.

Thanks to the scientific method persisting in pockets around the world, this gaping hole in the alarmists' argument does not go unfilled. As an experiment, Harvard-Smithsonian astrophysicist Dr. Willie Soon wondered if solar activity might have some interesting relationship with Arctic temperatures. His findings can be seen in the graphs at left.

Clearly, though potentially troubling to the CO_2-haters, the striking correlation between heat emitted by the sun and the measured temperatures in the Arctic must be a coincidence, because we have already concluded that it is George Bush and SUVs that are warming the Arctic (even if CO_2 emissions or GHG concentrations don't correlate with temperature either at present or historically). Or maybe our CO_2 emissions are heating up the sun.

Finally, after admitting that the *cause* and *extent* of Arctic warming are not quite what the media would have us believe, we ought to examine the *effect* of Arctic warming. *National Geographic* and the *New York Times* will tell the horror stories, but when the *San Francisco Chronicle* mentioned that gray whales were doing well and growing in population in the slightly warmer Arctic, the editors made sure to include the ominous "...for now."[6]

Sea no evil

Despite the apparent cuddliness of polar bears, the really compelling reason why a melting Artic might matter to us is that Al Gore tells us it will rise twenty feet and flood downtown Manhattan. His alarmist advisor James Hansen now talks of an eighty-foot rise in sea level. Now, if so many glaciers and ice caps have already melted as we are told, our celebrity hand-wringers should be able to notice the change from their back decks on Martha's Vineyard. Yet the beaches do look to be pretty much where we left them. Of course, the problems, regardless of the logic, are always just over the horizon (like those whales that "thrive in the Arctic, *for now*").

In *AIT's* book version, Gore presents page after page of *before-and-after* images depicting twenty feet of sea-level rise. All such images are as genuine as Gore's computerized bear that, unlike the real kind, can't swim. This rise would come from a purportedly impending collapse of the great ice sheets, for which no credible evidence exists, and further much of Greenland's ice were *Waterworld* ever to come about would actually

The Arctic Ocean is going down.

turn into a huge lake due to the topography. Regardless, were Gore actually interested, he could have referred to the data of NASA scientist Jay Zwally and colleagues on ice mass balance changes and potential effects on sea level.[7] Zwally, et al. (2005), found that the combined net loss of ice from Greenland/Antarctica would account for a sea-level rise equivalent of 0.05

Q: I gather from this last discussion that it would be absurd to attribute the Katrina disaster to global warming?

A: Yes, it would be absurd.

Kerry Emanuel of MIT, regularly cited by greens and media as claiming "the human did it!" on his website, http://wind.mit.edu/~emanuel/anthro2.htm

mm per year during 1992–2002. At that rate, it would take a mere millennium before sea levels shot up by a full five centimeters.

So, sea level rise is not a threat, except for those who might doze off for, say, twenty thousand years insisting all the while that their heads must remain at the tide line, are fortunate enough to escape a near-certain ice age, and who further defy history by refusing in the interim to respond or adapt at all to their circumstances—in this case a very, very slowly advancing tide. (This pretty well describes me on Spring Break while in college, if abbreviated because my folks would only spring for four years of tuition.)

We can be confident in a lax approach to Poseidon's wrath thanks to research from, among others, Nils-Axel Mörner of Stockholm University, who unlike the bulk of the IPCC's panel is in fact a recognized expert on sea levels. Mörner's research demonstrates that current sea levels are within the range of sea level oscillation over the past three hundred years, while the satellite data show virtually no rise over the past decade.[8] The most alarmist extant report, the IPCC Third Assessment Report, in its politician/bureaucrat/pressure group–drafted "Summary for Policymakers" no less, states the following about the past one hundred years during which the alarmists state that we have seen unprecedented melting worldwide: "Tide gauge data show that global average sea level rose between 0.1 and 0.2 metres during the twentieth century."[9] Later, again in the "SAP," it acknowledges that "Within present uncertainties, observations and models are both consistent with a lack of significant acceleration of sea level rise during the twentieth century."[10] This is the face-saving way of admitting that they can't claim that Man sped up the expected sea-level rise that occurs between glaciations (ice ages), which is the period in which we happily find ourselves.

The IPCC foresees sea-level rise of between fourteen and forty-four centimeters by 2100, not twenty feet as alarmists such as Al Gore bizarrely threaten, seemingly cut from whole cloth and certainly ignorant of past warmings. The Earth experienced a sea-level rise of twenty centimeters over the past century with no noticeable ill effects, given in large part because the rate of rise, throughout history, has been hardly noticeable and even if carried out over several centuries would certainly not be the stuff of Hollywood thrillers, however absurd.

In fact, Professor Mörner and his team visited the Maldives, which the IPCC says are at risk from sea-level rise (in slight conflict with the Maldives' expensive Brussels lobbyists, who, I'm told, seek millions from the EU to aid in developing ocean-front resort properties). Mörner found considerable evidence that the sea level around the islands has fallen over the past thirty years, and that the islands and their people had survived much higher sea levels in the past.

Remember, if floating ice melts, it has no effect on sea levels—only ice that melts from the land into the sea will raise waters. Basically, if warming is to raise sea levels, Greenland and Antarctica would be culprits. The charts right show Antarctica is not warming.

As noted, Zwally, et al., recently examined changes in ice mass "from elevation

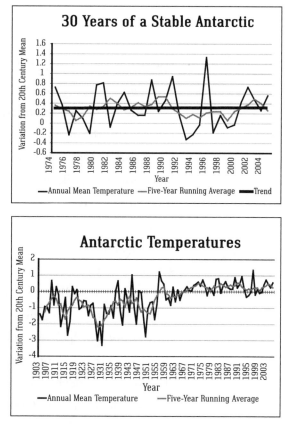

Source: NASA Goddard

changes derived from 10.5 years (Greenland) and 9 years (Antarctica) of satellite radar altimetry data from the European Remote-sensing Satellites ERS-1 and -2." The researchers report that the movements of these three ice sheets have added .05 to .03 millimeters to level of the sea per year. It is unclear where Al Gore gets his other 6095.95 millimeters of sea level to reach the 20 feet he warned of in his film.

The website www.CO2Science.org puts this sea-level rise in perspective by stating that: "At the current sea-level-equivalent ice-loss rate of 0.05 millimeters per year, it would take a full millennium to raise global sea level by just 5 cm, and it would take fully 20,000 years to raise it a single meter."

Sleep well. But in the year 22007, do *not* hit that snooze button. Or else stand up, brush off your Hawaiian shorts, and hitch a ride back to class.

Doing something about the weather

If the glacially inspired rising seas don't get you wet, the alarmists might say, the CO_2-caused hurricanes will. Hurricanes purportedly will increase with an increase in sea surface temperatures caused by the build-up of Manmade GHGs. Curiously, the oceans have been cooling of late (though despite seemingly upturning the apple cart of alarmism, it is incoherently dismissed instead as the gathering storm merely taking a breather).[11]

Cries of more frequent and severe weather caused by Man are an alarmist staple. The trouble is that the weather isn't cooperating, and the research never has. Regarding the former, a calm 2006 hurricane season left the alarmists twisting in amusing ways to distract attention from the absence of calamity they had confidently promised only several short months prior. Regarding the latter, for example, research by German scientists has demonstrated that the devastating floods in central Europe in 2002 were perfectly normal when compared against the historical record over centuries.[12] This may not be as authoritative as re-election campaign rhetoric to the contrary

Statement on the U.S. Hurricane Problem, July 25, 2006

As the Atlantic hurricane season gets under way, the possible influence of climate change on hurricane activity is receiving renewed attention. While the debate on this issue is of considerable scientific and societal interest and concern, it should in no event detract from **the main hurricane problem facing the United States: the ever-growing concentration of population and wealth in vulnerable coastal regions.** These demographic trends are setting us up for rapidly increasing human and economic losses from hurricane disasters, especially in this era of heightened activity. Scores of scientists and engineers had warned of the threat to New Orleans long before climate change was seriously considered, and a Katrina-like storm or worse was (and is) inevitable even in a stable climate.

Rapidly escalating hurricane damage in recent decades owes much to government policies that serve to subsidize risk. State regulation of insurance is captive to political pressures that hold down premiums in risky coastal areas at the expense of higher premiums in less risky places. Federal flood insurance programs likewise undercharge property owners in vulnerable areas. Federal disaster policies, while providing obvious humanitarian benefits, also serve to promote risky behavior in the long run.

We are optimistic that continued research will eventually resolve much of the current controversy over the effect of climate change on hurricanes. But the more urgent problem of our lemming-like march to the sea requires immediate and sustained attention. **We call upon leaders of government and industry to undertake a comprehensive evaluation of building practices, and insurance, land use, and disaster relief policies that currently serve to promote an ever-increasing vulnerability to hurricanes.**

> **Kerry Emanuel, Richard Anthes, Judith Curry, James Elsner, Greg Holland, Phil Klotzbach, Tom Knutson, Chris Landsea, Max Mayfield, Peter Webster**

http://wind.mit.edu/~emanuel/Hurricane_threat.htm (emphases added)

by Gerhard Schröder, but even in Germany, they couldn't quite pull off blaming America for storms back in the days when Germans were burning "witches" over climate change and America had yet to be born.

The most recent *"an expert agrees with me!"* poster-child for hurricanes is MIT professor Kerry Emanuel, who was ritually cited for a paper of his that argued for a clear correlation between climate change and hurricanes. As man warms the oceans, the argument goes, hurricanes intensify (and cause greater damage, again ignoring the obvious development-related factors for increased storm losses). Christopher Landsea of the National Oceanic and Atmospheric Administration (NOAA), who found no trend for land-falling U.S. hurricanes, suggests that Emanuel's finding may be an "artifact of the data"—a consequence of improved satellite detection, monitoring, and analysis of non-land-falling hurricanes. Also, the evidence of cooling seas ought to cool the rhetoric about the active 2005 season. Instead, it is alternately disputed or ignored: disputed by the alarmists, and ignored by the media until the dispute turns into something more tangible.

Regardless, Emanuel's subsequent public utterances certainly belie the originally flogged conclusion. A calmer claim that recent storms were mostly bad luck, made amid a calmer 2006 storm season, somehow missed the widespread coverage. "The high-impact hurricanes that have hit the United States over the past couple years are, at least for now, more a function of bad luck than of climate change, said MIT Professor Kerry Emanuel during an October 31 [2006] symposium . . . .' On a fifty-year time scale from a U.S. point of view, it probably doesn't mean anything at all,' he said. Only about one third of the storms over the Atlantic even make landfall. 'The last two years have been more or less bad luck,' he said."[13]

This acknowledgment, not widely reported, was run with the amusing headline: "Hurricane horror mostly bad luck—for now." Note not just the de rigueur "horror" but the gratuitous "For now," as the all-knowing press service ensured its own views entirely swamped Professor Emanuel's musings.

The numbers game

The weather has of late been more damaging, we are told, which historically would actually be unremarkable. Man's activity has *made* the weather more damaging, they say. Well, agreed, but not as they'd like one to believe. Whereas the greens will say we do this by driving too much, common sense reminds us that people increasingly develop and occupy storm-prone areas. We build homes on flood plains, and beach getaways on the Outer Banks of North Carolina (you can find quite a few politicians and lobbyists here

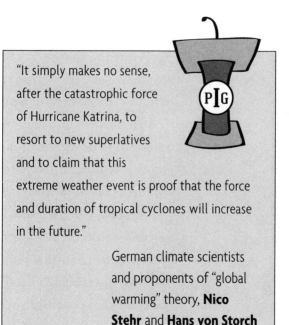

"It simply makes no sense, after the catastrophic force of Hurricane Katrina, to resort to new superlatives and to claim that this extreme weather event is proof that the force and duration of tropical cyclones will increase in the future."

German climate scientists and proponents of "global warming" theory, **Nico Stehr** and **Hans von Storch**

during the August congressional recess). Few to none of these would be built without taxpayer subsidy in the form of guarantees bringing insurance costs down from what would be astronomical rates, if the insurance was available at all. The more we build in areas certain to flood or get walloped by a hurricane, the more costs will rise annually.

Presumably, up next is a federal guarantee against loss for those who build atop railroad tracks, protecting against the equally freak chance that they get plowed over by the 7:12 to Peoria. (Of course, in this case the green lobby would ritually tout the inexplicable and mounting losses to advance their campaign against the railroads which, after all, deliver half our nation's energy supply. *"Coal use proven to destroy homes. Send money to help us to continue our struggle against this Black Death!"*)

Construction of course follows population. On this matter, alarmists such as Al Gore fail to acknowledge that a more populous world, with more residents in more areas that with some regularity experience rising waters, has more events rising to the level (so to speak) of being considered "floods."

While the tree falling in an empty forest remains an unsettled question, if water arrives and there's nothing to wash out then, no, the "flood" *doesn't* really happen. High tide is not a flood unless you built a housing development there during low tide. When we build things or place lives or wealth in more places that might get covered with water, more storms and floods get reported, and more damage is claimed.

Just as people are changing where they live and put property, our observation and reporting technologies are improving—again yielding more recorded storms.[14] As regards tornadoes, consider the advances in Doppler radar. A tornado that would have gone undetected in the past (either because it was smaller than old technology could catch, or because nobody on the ground actually saw it or was hit by it), gets recorded today. We not only get hit by more of them when there are more of us in more places, but with continuing improvements in this radar, we can detect smaller and smaller twisters. Thus, we have more tornadoes on record.

Hurricane Kyoto

During hurricane season, Pat Robertson suddenly gains crunchy allies by saying the deadly storms are the consequence of human misbehavior. The mania of blaming modernity for hurricanes hit fever pitch with the long-expected return in 2005 of more frequent storms, consistent with standard forty- to fifty-year cycles. These predictions and well-established historical cycles did nothing to impede claims that the 2005 hurricane season was exacerbated by global warming, any more than the almost nonexistent 2006 storm season led the alarmists to retract their guarantees of calamity. In response to this sensational abuse one top expert, Chris Landsea of NOAA resigned in protest from the UN's IPCC in January 2005, detailing precisely how and why such claims, by a purportedly scientific body turned wildly political, were fabrications. In so doing he joined other world-class experts like Lindzen in giving up on the IPCC's hopelessly political, non-scientific project. Amazingly, there was little to no fallout and the Man-causes-hurricanes storyline continued unabated as the 2006 season kicked off.

More Hurricanes=Global Warming; Fewer Hurricanes=Not News

This wire story got a lot less notice than the 2005 declarations that more hurricanes proved global warming's destructive power.

Atlantic hurricane season about to end, with no U.S. landfall

MIAMI (AFP) — Residents of the southern USA heaved a sigh of relief as a comparatively quiet Atlantic hurricane season nears its conclusion with none of the storms making U.S. landfall.

"The 2006 Atlantic basin hurricane season was much less active than the 2004 and 2005 seasons, but 2006 was also atypical in that there were no land-falling hurricanes along the U.S. coastline this year," leading hurricane expert William Gray said in a report released on Friday.

"This is the first year that there have been no landfalling hurricanes along the U.S. coastline since 2001, and this is only the 11th year since 1945 that there have been no U.S. landfalling hurricanes," said Gray, of Colorado State University.

So far in the six-month season that ends on November 30, there have been nine named storms, the lowest number since 1997. Five of those storms developed in hurricanes, two of which were considered major.

Agence France-Press, November 20, 2006

The hurricane season indeed did end without event a week later.

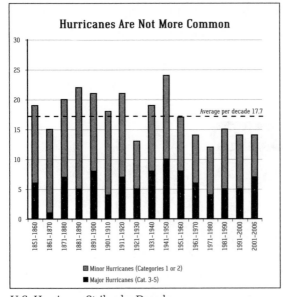

Hurricanes Are Not More Common

Average per decade 17.7

Minor Hurricanes (Categories 1 or 2)
Major Hurricanes (Cat. 3-5)

U.S. Hurricane Strikes by Decade
(Source: National Weather Service)

Stanley Goldenberg of NOAA's Hurricane Research Division plainly states that "Katrina is part of a well-documented, multi-decadal scale fluctuation in hurricane activity. This cycle was described in a heavily cited article printed in the journal *Science* in 2001....I speak for many hurricane climate researchers in saying such claims are nonsense."[15]

Bad hurricane seasons come in cycles. NOAA's Landsea agrees with Goldenberg, saying: "If you look at the raw hurricane data itself, there is no global warming signal. What we see instead is a strong cycling of activity. There are periods of twenty-five to forty years where it's very busy and then periods of twenty-five to forty years when it's very quiet."

Possibly because real experts won't play along, the American Geophysical Union released a paper in late June 2006 (discussed elsewhere in these pages) concluding in essence that hurricane experts must not know what they're talking about.

Data from NOAA's U.S. National Hurricane Center[16] demonstrates that the most active hurricane decade on record was 1941–1950. That decade saw twenty-four hurricanes hit the U.S., of which ten fell into the large storm Categories 3, 4, and 5. Those major storms were most frequent in the 1890s, 1930s, and 1940s, not now. Two-thirds of the ninety-two large hurricanes striking U.S. shores between 1851 and 2004 occurred before 1950.

These analyses come from a NOAA report, "The Deadliest, Costliest and Most Intense United States Tropical Cyclones from 1851 to 2004."[17] The report's Table 6 lists hurricanes by decades since 1851 and shows that during the forty-year period 1961–2000 both the number and intensity of landfalling U.S. hurricanes *decreased* sharply. It was updated in July 2006, and the data for 2005, when frequency was high, demonstrating that, at six, the number of hurricanes affecting the continental U.S. in 2005 matched 2004's total as well as that of 1886, 1916, and 1985, and was one more than numerous other years.[18] Winds that year ranged from sixty-five knots (Cindy and Ophelia) to 110 knots (Katrina). Compare this to at least eight storms prior to Katrina with more powerful winds—possibly in the mid-teens as a handful, measured with less sophisticated equipment, registered at the same strength—and no wind speeds are recorded at all for three and a half decades, from the 1940s to the 1970s.

If 2005 marked a trend then so did the calm 2006, indicating that Man's impact has swung yet again. Just as we went from causing global cooling to global warming, now even more GHGs in the atmosphere must, as a matter of logical consistency, have ensured a calm storm season. Or maybe the weather's only your fault when it's bad.

Global warming causing glaciers to grow: study

"A new study has found that warmer winters and cooler summers, bringing increased precipitation, could be causing some glaciers to increase in size.

Newcastle University researchers found that the western Himalayas' Upper Indus Basin was experiencing more snow and rainfall, which has implications for the water supplies of about 50 million people in Pakistan.

The study appears in the American Meteorological Society's Journal of Climate (BBC News online)."

August 25, 2006, report from Greenwire, unable to utter what kind of "implications" such increased precipitation and cooler summers might have on Pakistan's millions.

The more comprehensive 2005 version of NOAA's report affirms data showing that the world *cooled* from 1940 to 1975. We know that warming—allegedly Manmade—has occurred since about 1975, or just before the "consensus" switched, unannounced, from a looming ice age to an inevitable burning, both being Man's fault. Therefore, the post-1975 storm data is most relevant. Recall the lore that the 1990s were a historically unprecedented decade of heat.

The most intense hurricane to hit the U.S. was still in 1935, the "Labor Day" storm (human names were not affixed until 1950). During this storm, the barometer, measuring air pressure, fell to 892 millimeters (the lower the pressure, the *more* intense the storm). The second most intense storm was in 1969 (minimum pressure 909 mm). Only the third most intense was after 1975, and that was in 1992 (at 922 mm, Hurricane Andrew was much less intense than the 1935 Labor Day hurricane, which also hit Florida, but Andrew was far more costly, for obvious reasons discussed above). The report arguably reveals that the intensity of Category 5 hurricanes has *decreased* since 1935. It could further be argued that the intensity of Category 4 hurricanes has decreased since 1886. It is equally arguable from the data that a decrease in the degree *and* frequency of hurricanes has been seen since the 1940s, both the more severe Category 3–5 hurricanes and also total storms, Category 1–5. Yet we need not bother playing the greens' baseline game to make that argument, acknowledging instead the forty- to fifty-year storm cycle.

Whatever the data tell us, clearly these storms striking the U.S. since the Industrial Revolution have been and remain cyclical. Period. Alarmists are left playing the "well, what if..." game.

Let us look even further. There is no overall *global* trend of hurricane-force storms getting stronger. Throughout most of the planet, storms have not shown any noticeable change. In two areas of the planet storm strength has shown a statistically significant change: in the North Atlantic, hurricanes are getting stronger, and in the North Pacific cyclones are getting *weaker* (it is only a matter of time before this merciful trend in the Pacific is declared a disastrous consequence of Manmade global warming; until then it will continue to be ignored).

Climate scientist Roger Pielke, Jr., addresses this specifically in the context of claims of increased storm losses, summarizing the IPCC Third Assessment Report's conclusions on the relevant matters:

> Losses have continued to increase, and the IPCC still has not identified any secular trends in weather extremes, with only one exception. The IPCC found no long-term global trends in tropical or extra-tropical cyclones (i.e., hurricanes or winter storms), in "droughts or wet spells," or in "tornados, hail, and other severe weather."[19]

In 2006, the Bulletin of the American Meteorological Society published a paper by an interdisciplinary team of experts.[20] Their three main points were:

1) there is no established connection between greenhouse gas emissions and the observed behavior of hurricanes;
2) any future changes in hurricane intensities will likely be small and within the context of observed natural variability; and
3) the politics of linking hurricanes to global warming threatens to undermine support for legitimate climate research and could result in ineffective hurricane policies.

Other recent studies also cast extreme doubt on the media claims about warming's influence on hurricanes. P. J. Klotzbach[21] finds that: "*The data indicate a large increasing trend in tropical cyclone intensity and longevity for the North Atlantic basin and a considerable decreasing trend for the Northeast Pacific. All other basins showed small trends, and there has been no significant change in global net tropical cyclone activity.*

There has been a small increase in global Category 4–5 hurricanes from the period 1986–1995 to the period 1996–2005. *Most of this increase is likely due to improved observational technology.* These findings indicate that other important factors govern intensity and frequency of tropical cyclones besides SSTs [sea surface temperatures]" (emphases added).

Further, in July 2006, *Science* magazine actually let an article slip through its sieve identifying a serious data problem with the nascent studies attempting to link hurricanes and Manmade global warming.[22]

False Prophecies

"2006 hurricane forecast: 8–10 storms

The 2006 Atlantic hurricane season will be very active with up to 10 hurricanes, although not as busy as record-breaking 2005, when Hurricane Katrina and several other monster storms slammed into the United States, the U.S. government's top climate agency said on Monday.

'NOAA is predicting 13 to 16 named storms, with eight to 10 becoming hurricanes, of which four to six could become "major" hurricanes of Category 3 strength or higher,' said Conrad Lautenbacher, administrator of the National Oceanic and Atmospheric Administration.

Some climatologists, however, say there are indications that human-induced global warming could be increasing the average intensity of tropical cyclones."

MSNBC, May 22, 2006

"Hurricane Season Blows Past, Quietly

Instead, it has been a long, lazy hurricane season with just half the number of hurricanes predicted and not a single one making landfall."

ABC News, October 24, 2006

Strangely, the Associated Press covered this paper and with eerie fairness.[23] In brief, the researchers found that outdated technology ensured that the severity of previous storms was underestimated, thereby enabling two recent claims of increasing hurricane intensity, which was naturally then pinned on Man.

Certainly, as each new hurricane season approaches and particularly in the Atlantic basin, we will increasingly hear about this *über*-important wind and rain, given that it not only lashes Washington, D.C., but the epicenter of the media world, New York City. Helpfully, even *New York* magazine weighed in to calm its readers before the media storm, reminding them in the summer of 2006 that Manhattan has been beaten about by a "monster storm" about every seventy-five years, though current density and development means the *impact* becomes over time more severe.[24]

Don't count on such media sensibility being widespread once the rain falls.

Drought

If science gets in the way of alarmists claims that global warming–induced rising oceans and falling skies will soak you, then surely global warming will drain you dry—and of course Man is the culprit. If not storms, then droughts must *certainly* be on the rise, again in frequency and severity. Yet once again the only detectable increase is in news reports claiming an increase.

In late September 2006, Al Gore repeated an annual media mantra in his alarmist speech at New York University: "Warmer temperatures have dried out soils and vegetation. All these findings come at the end of a summer with record breaking temperatures and the hottest twelve-month period ever measured in the U.S., with persistent drought in vast areas of our country."

Time magazine's absurd "worried" cover story even elevated drought to cover-boy status: "Polar Ice Caps Are Melting Faster Than Ever . . . . More and More; Land Is Being Devastated by Drought....Rising Waters Are Drowning Low-Lying Communities," and so on. The *Washington Post* ran a front-page piece whose thesis was a none-too-vague threat that Africa would be beset by swarms as a result of Manmade drought: not of locusts, but trial lawyers looking to blame your profligate energy use.[25]

Yet consider actual research, if one may be so bold as to suggest that heresy: "An increasing trend is apparent in both model soil moisture and runoff over much of the U.S., with a few decreasing trends in parts of the Southwest. The trend patterns were qualitatively similar to those found in streamflow records observed at a station network minimally affected by anthropogenic activities. This wetting trend is consistent with the general increase in precipitation in the latter half of the twentieth century. *Droughts have, for the most part, become shorter, less frequent, and cover a smaller portion of the country over the last century.*" Andreadis and Lettenmaier, "*Trends in twentieth-century drought over the continental United States*" (emphasis added).[26] Even the IPCC could not corroborate Mr. Gore's panic.[27]

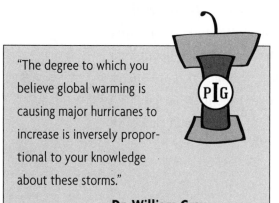

"The degree to which you believe global warming is causing major hurricanes to increase is inversely proportional to your knowledge about these storms."

Dr. William Gray, Hurricane Research Guru

Ah, those pointy-heads probably drive SUVs and have sold out anyway.

Al Gore: Couldn't stand the weather

Naturally, *An Inconvenient Truth* claims an unnatural increase in floods and droughts as well as desertification. With warmer air there tends to be more water vapor, which means more precipitation. This helps explain why amid the slight warming of past decades the Sahara Desert has, since the mid-1980s, begun to *recede* at its southern borders after several decades of advance, contrary to Gore's stunning photos.[28] Even *New Sceintist* magazine shrieked that "Africa's deserts are in retreat." Shrinking deserts—meaning thriving plant life—are actually a signal of global warming, but Gore pretends the opposite is happening, just because it's scarier.

As discussed above, the same amount of flooding will yield more damage if more people live on the flood plains. It's an important point, and one that Al Gore conceals. Consider this passage from Gore's book of *An Inconvenient Truth:* "Partly as a result [of climate-related hydrologic changes], the number of large flood events has increased by decade by decade, on every continent."[29] It's a convenient claim for Gore, but the source he invokes was actually quite clear that it is virtually impossible to tease out warming as a cause of increased flood damage because of the intensely complex commingling of potential factors—"noise"—created by natural climate variability and socioeconomic changes.

That some players in the reinsurance industry—got that, "*industry*"?—claim a fear of Manmade future disasters as a reason to raises their rates and beg for government guarantees is often cited by alarmists as further proof that their predicted doom must be believed. Yet, consider scientist Roger Pielke, Jr., pointing out several problems with this, beyond the industry's conflict in serving as such a backstop for the purportedly

> "If long-term trends are accepted as a valid measure of climate change, then the air temperature and ice data do not support the proposed polar amplification of [CO_2-induced] global warming. The potential importance of large-amplitude variability and numerous feedbacks involved in Arctic atmosphere-ice-ocean interactions implies that the Arctic poses severe challenges to generating credible model-based projections of climate change."
>
> **Polyakov, Akasofu** and seven other colleagues from the International Arctic Research Center (2002, *Eos*, vol. 83, no. 47, 547–548)

authoritative IPCC—"the United Nations, home to the venerable IPCC, and advocacy groups often partner with reinsurance experts to advance their agenda. Not only does this not make sense for intellectual reasons, as the UN's IPCC is supposed to be the authority of climate science (why do they need reinsurance industry backup?) but also for pragmatic reasons. . . . When the reinsurance industry makes claims of increasing disasters, from the perspective of conflict of interest, this is no different than a fossil fuel interest promoting that science that best supports their interests."[30]

Al Gore, however, uses a chart to push his claim, presuming as always that the reader will simply be too busy curled in the fetal position to check Gore's work. This particular chart shows that the number of major floods in, for example, Asia, increased from under 50 in 1950–1959 up to more than 300 in 1990–2000.[31] Gore indicates that his chart simply measures changes in the number of major floods, *i.e., a change in the number of physical events.* The only interpretation of this, in Gore's view, is that Man causes severe weather. (After all, while ultimately a polemic against population, *AIT* is hardly a movie about the perils of increased construction in potential flood zones.)

But as Gore's own source makes clear, the chart actually measures changes in the number of "*damaging*" floods. Are these the actual, physical event? Not at all. The source, the Emergency Disasters Database, explains: "Only events that are classified as disasters are reported in this database. (An event is declared a disaster if it meets at least one of the following criteria: ten or more people reported killed; one hundred or more people reported affected; international assistance was called; or a state of emergency was declared.)"[32]

Of course, then, the database Gore uses is going to be skewed toward more events in later decades, offering the appearance of increased flooding. Gore blames fossil fuels and warming. The real "culprits" are superior recordkeeping, more extensive insurance coverage, more declared states of emergency, more people, more development, and more official calls for international assistance. The very text that Gore refers to but, again and for obvious reasons does not cite, acknowledges this: "Figure 16.5 shows a clear increase in the number of floods since the 1940s for every continent *and a roughly constant rate of increase for each decade. However*, it should be noted that although the number has been increasing, *the actual reporting and recording of floods have also increased since 1940, due to the improvements in telecommunications and improved coverage of global information*"[33] (emphases added).

The text upon which Gore relies goes on to point out:

> "[T]he often predicted impact of climate change has become a reality in that poor sections of society living in coastal regions bore the brunt of the hurricane."
>
> Draft Motion, **European Parliament**, September 5, 2005, blaming Man, implicitly *Homo Americanus*, for Hurricane Katrina.
>
> (So, before "Manmade global warming," the poor living on coastal regions stared happily overhead as storms proceeded on to devastate rich, inland areas.)

> Flood processes are controlled by many factors, climate being one of them. Other non-climatic factors include changes in terrestrial systems (that is, hydrological and ecological systems [such as wetlands loss and deforestation]) and socioeconomic systems. In Germany, for instance, flood hazards have increased . . . partly as a result of changes in engineering practices, agricultural intensification, and urbanization.[34]

In short, just as we get "increased storm damage" when people build more in storm-prone areas, the number of "damaging" floods or floods classified as "disasters" also increases with population growth and development in flood plains.

Gore, like everyone touting increased storms, floods, melting things, and sea-level rise, is simply trying to scare you by fooling you.

Part III

* * * * * *

THE FALSE PROPHETS (AND REAL PROFITS) OF GLOBAL WARMING

Chapter Eight

✳ ✳ ✳ ✳ ✳ ✳

MEDIA MANIA
GOOD NEWS IS NO NEWS

"There are ominous signs that the Earth's weather patterns have begun to change dramatically," *Newsweek* warns, "and that these changes may portend a drastic decline in food production—with serious political implications for just about every nation on earth. The drop in food production could begin quite soon....The evidence in support of these predictions has now begun to accumulate so massively that meteorologists are hard-pressed to keep up with it."

The year was 1975, and the threat was global cooling, about which the magazine said a year later, "this trend will reduce agricultural productivity for the rest of the century."

The need to sell copy or attract viewers, together with the general government-as-savior leanings and resentment of real businessmen, drive the media to embrace—in fact *drive*—environmental and climate alarmism. Alarmism of any stripe (sharks, bird flu, gas prices, unemployment, nuclear war) is a media staple, but environmentalism provides them some of their greatest delights.

American Enterprise Institute visiting scholar Joel Schwartz compiled some of the greatest (worst?) headline displays of what he calls journalism's "mother's milk of gloom and doom":

"Air Pollution's Threat Proving Worse Than Believed."

"Don't Breathe Deeply."

Guess what?

✳ The media have alternated between global cooling scares and global warming for a century.

✳ Many journalists proclaim a duty to present only one side in climate debates.

✳ The media ignore evidence that would deflate global warming fears, and exaggerate that which cuts "their way."

> ❄❄❄❄❄❄❄❄❄
>
> "Despite all you may have read, heard, or imagined, it's been growing cooler—not warmer—since the Thirties," stated a seemingly prescient **Fortune** magazine in "Climate—the Heat May be Off," 1954.

"Study Finds Smog Raises Death Rate."

"State's Air Is Among Nation's Most Toxic."

"Asthma Risk for Children Soars with High Ozone Levels."

Schwartz writes: "Headlines like these might be warranted if they accurately reflected the weight of the scientific evidence. But they do not. Through exaggeration, omission of contrary evidence, and lack of context, regulators, activists, and even many health scientists misrepresent the results of air pollution health studies and the overall weight of the evidence from the research literature. They create the appearance that harm from air pollution is much greater and more certain than suggested by the underlying evidence."[1]

Pressure groups of course provide the fodder for the lazy wretches of environmental journalism, issuing such stereotypically titled screeds as "Danger in the Air," "Death, Disease, and Dirty Power," "Highway Health Hazards," "Plagued by Pollution," and "Children at Risk." Schwartz notes that even health researchers play to the media's love of alarmist storylines, grabbing headlines with press releases blaring "Smog May Cause Lifelong Lung Deficits," "Link Strengthened between Lung Cancer, Heart Deaths and Tiny Particles of Soot," "USC Study Shows Air Pollution May Trigger Asthma in Young Athletes," and "Traffic Exhaust Poisons Home Air." Gasp!

Something about frozen places, however, truly drives the media out of their collective minds such that "global warming" reportage casts the ink-stained mob's air pollution hysteria as a relative paragon of reportorial virtue. No "climate" story is implausible. The Great Alaskan Warming warned of by the *New York Times*, among others, turns out to be no more than one major warming occurring in 1976–1977, likely a result of what climate science calls the Pacific Decadal Oscillation (the *Times* was never able to identify a source for its claim, and printed a retraction, sort of). Yet news stories and even *West Wing* story lines continue to proliferate about "Baked Alaska" with no substantive underpinning.[2]

Things get comical in the rush to proclaim horror not just in a future hot planet as the alarmists envision, but that which we experience today (warmer relative to the Little Ice Age of not much more than a century ago, that is). Consider the *Times*'s science correspondent Andrew Revkin, generally more sober than his peers, writing the following on Arctic warming and dragonflies:

> The global warming trend that raised the earth's average temperature one degree Fahrenheit in the twentieth century has had a stronger effect here [in the Arctic]. . . . The planet has for millions of years seen great cycles of ice ages and warm periods . . . .But each year brings more signs that recent environmental shifts around the Arctic are extraordinary. *Dragonflies are showing up for the first time in memory in Eskimo villages, causing children to run to their parents, scared of these unfamiliar insects*[3] (emphasis added).

There are of course several problems with the thesis, including the unmentioned one-time contribution of the Pacific Decadal Oscillation that, lamentably, will likely soon be reversed. There is also the fact that even larger and more rapid warming in the Arctic was observed at least in the 1910s–1920s with no plausible connection to CO_2 or greenhouse warming.

Regarding the claim that dragonflies have suddenly appeared, brought northward by the suddenly temperate climes,

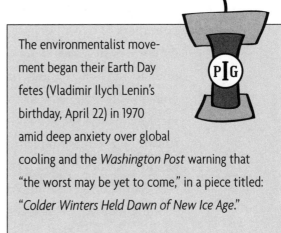

The environmentalist movement began their Earth Day fetes (Vladimir Ilych Lenin's birthday, April 22) in 1970 amid deep anxiety over global cooling and the *Washington Post* warning that "the worst may be yet to come," in a piece titled: *"Colder Winters Held Dawn of New Ice Age."*

At the Movies

"You know, when I was a kid, food was food! Until our scientists polluted the soil…decimated plant and animal life. Why, you could buy meat anywhere. Eggs, they had. Real butter. Fresh fruit in the stores! How can anything survive in a climate like this? A heat wave all year long! The greenhouse effect! Everything is burning up!"

Sol Roth (Edward G. Robinson), *Soylent Green* (1973)

the bug was known to proliferate sufficiently that *thirty-one species of dragonflies* (with larval life span lasting as long as a decade in alpine and high-latitude habitats) are known in Alaska and "relatively abundant," according to that state's bugmen.[4] At least fifteen years ago the four-spot skimmer dragonfly was pervasive enough such that it was voted by (terrified, no doubt) elementary school children in Alaska to be the state insect, out-polling the mosquito, butterfly, and bumblebee (all of which we doubtless will soon hear have also just arrived).

Such are the perils of journalism- or science-by-anecdote. This might not rival the claim that Inuits are having their millennia-long traditional way of life disrupted by, for example, having their snowmobiles fall through the ice. Verily, we admit, this did not befall them in ancient times. I have also heard the Inuits' non-Inuit lawyer relate the crushing disappointment of learning, upon moving up to share their lifestyle, that they do not live in igloos as he was expecting, but climate-controlled homes with all the electric modern conveniences. Why these and other indigenous people won't stay like Western elites love to imagine them is beyond me. Meanwhile the Inuits themselves also complain about their airport's runway buckling from summer warming, and the high cost of gasoline. Oh, for the sacred days when a gentler Mankind did not have these horrors visited upon his indigenous peoples by a vengeful Gaia!

But the dearth of perspective in media coverage goes further north than Alaska.

ABCs of Global Warming:
Cold=Weather, Hot=Climate

August of 2006 saw a three-week heat spell, and ABC declared it "new proof of global warming." Here's the transcript, courtesy of Media Research Center:

Bill Blakemore: Yes. Chris, it's been twenty-one days of brutal heat, a coast-to-coast double heat wave leaving at least 186 people dead, massive damage to crops and livestock. And yes, it's got folks everywhere asking if it's part of global warming, and thus a sign that Earth will keep getting hotter.

Bill Blakemore: (Voiceover) The scientists say yes, global warming is involved. First, it fits the pattern predicted thirty years ago, more frequent and intense heat waves. More than fifty cities just broke records.

☀-☀-☀

Just three months later, after an un-newsworthy hurricane season and a two-month cold spell, ABC News put things in perspective:

You probably noticed there were fewer Atlantic hurricanes this year. Melting Arctic sea ice came extremely close to but didn't break the record minimum of summer 2005. And today, the National Oceanic and Atmospheric Administration, or NOAA, announced two months of cooler-than-average temperatures across the United States.

So what happened to global warming?

Scientists who study climate say they get that question every time there's a cold spell. Their answer: It's important to keep in mind an important concept—weather is not climate.

All the alarmism that's fit to print

On the night of August 18, 2000, Al Gore, who had built his career around environmental issues, accepted the nomination of the Democratic Party. In his speech, he declared, to much applause, "We must reverse the silent rising tide of global warming, and we can."

"You'll find more dissent at a North Korean political rally than in this program."

Dave Shifflett review for Bloomberg News of Tom Brokaw's "global warming" special, July 16, 2006

The next day, the *New York Times* (just below its unseasonably cool forecast of 62 degrees overnight) carried on Page 1 a smiling, waving shot of Al Gore and his wife Tipper. Alongside the photo, the *Times* also ran this Sunday cover story: "Ages-Old Icecap at North Pole Is Now Liquid, Scientists Find."

The first sentence was dramatic and breathless: "The North Pole is melting." The article continued: "The thick ice that has for ages covered the Arctic Ocean at the pole has turned to water... something that has presumably never before been seen by humans and is more evidence that global warming may be real and already affecting climate."

The source of this startling report was professor Malcolm McKenna, who drove home the point: "Some folks who pooh-pooh global warming might wake up if shown that even the pole is beginning to melt at least sometimes."

This was a smoking gun for the *New York Times*, and they planned to wave it around quite a bit. Four days later, on their op-ed page, the *Times* carried a piece by Edmund Blair Bolles, declaring "The nineteenth century's dream of an open polar sea has become the twenty-first century's nightmare."[5] Bolles declared that the water up north carried "potentially dire implications for the environment." He wrapped up his alarmist piece by declaring, "For the first time in 50 million years, we now have an open polar sea."

Then on August 28, as part of its series on Campaign 2000, the *Times*'s lead editorial titled "Protecting the Earth" began by repeating "Earlier this month, scientists spotted a patch of open ocean about a mile wide at the North Pole," and ended by declaring Al Gore a better candidate for the environment.

The next day saw the *Times*'s fourth A-section treatment of this melted-North-Pole story. It was an op-ed by Gregg Easterbrook that began, "That North Pole ice has turned liquid may be the least of our problems." The same day, however, if you looked in a corner of the

Idle Hands

As August 2006 petered out and the hurricane season plodded along without helpful storms to shriek about—a troubling development that continued into "the season"—green pressure groups issued daily press releases commemorating the one-year anniversary of Hurricane Katrina with the ultimate low in the debating dodge of an appeals to authority, by listing politicians and pastors who accept alarmism. *How crazy can I be? Pat Robertson believes me!*

Equally confused, MSNBC blared the headline: "Hurricane chief: Megadisaster 'is coming'"

The actual quote by Max Mayfield, director of the U.S. National Hurricane Center, was that it *could be centuries off* and was only if we keep asking for it, not by using energy, but *by building huge targets in storm-prone areas*:

> "I don't know whether that's going to be this year or five years from now or a hundred years from now. But as long as we continue to develop the coastline like we are, we're setting up for disaster."

News Summary page, you would see small box titled "Corrections." It's worth running in its entirety:

> A front-page article on Aug. 19 and a brief report on the Aug. 20 in *The Week in Review* about the sighting of open water at the North Pole misstated the normal conditions of sea ice there. A clear spot has probably opened at the pole before, scientists say, because about 10 percent of the Arctic Ocean is clear of ice in a typical summer.
>
> The reports also referred incompletely to the link between the open water and global warming. The lack of ice at the pole is not necessarily related to global warming.

The *Times* covered this misstep in more depth on page F-3 that day, just the page after the story about new ways of cooking with salt. In contrast to the A-1 drama from nine days earlier, this back-section article, titled "Open Water at Pole Not Surprising, Experts Say," included un-exciting quotations from actual climatologists such as "there's nothing to be necessarily alarmed about. There's been open water at the pole before. We have no clear evidence at this point that this is related to global climate change."

In 2006, when icebreakers encountered unusually thick ice down in Antarctica, however, this was unfit to print.

Full court press

The press and entertainment elites' global warming breathlessness culminated in a massive, combined-forces and multi-media effort kicking off in 2006. Those panting media voices that have regularly touted both the cooling and warming scares—oddly, without reporting the reversal of humanity's fate, let alone explaining their own flip-floppery—now actually question whether the public's failure to clamor for radical proposals is proof that the "news" has been "too balanced." In July 2006 the alarmist web-blog RealClimate.org introduced "RC Forum," representing a relaxation of their supposed science-only focus to allow more space for venting. The first entry was about how climate journalism is better in Europe because it doesn't bother with the pesky idea of "balance."[6]

The fear of balance is particularly acute now that powerful forces are doubtless coordinating such resistance. Columnist George Will noted, "An article on ABC's web site wonders ominously, 'Was Confusion over Global Warming a Con Job?' It suggests there has been a misinformation campaign implying that scientists might not be unanimous."[7] Paraphrased: *How can these rubes still care about the scientific dissent? Don't they know that we disapprove?*

CBS's Scott Pelley revealed that allowing balance in "global warming" stories is unforgivable bias. In the context of defending his laughably alarmist spring 2006 package for *60 Minutes*, Pelley let slip that "There becomes a point in journalism where striving for balance becomes irresponsible."[8] That point is apparently *now*, as in the same interview Pelley analogized climate change "skeptics" to Holocaust deniers.

No News There

"[T]he story [of debunking the global warming icon, the 'Hockey Stick'] has been largely ignored by Canadian media. The *Globe and Mail* has yet to carry one story on the subject. Only a handful of scattered references have appeared in other newspapers across the country, notably the *Calgary Herald*. The *Toronto Star*'s only acknowledgement was a column by Jay Ingram, who essentially said the debate among scientists was undermining public confidence in science. Two Canadian scientists rock the world climate community, trigger international reaction, congressional investigation and a National Academy of Sciences report that supports the Canadians' critique of official United Nations science. No news there, apparently."

Terence Corcoran, "See the Truth on Climate History,"
National Post, July 12, 2006

Desperately seeking to speak in terms understood by the unwashed in order to convert them, as the annual catastrophe of summer warming approached the Northern Hemisphere in 2006 *ABC News* posted a casting call for individuals to email their own heart-wrenching stories of how global warming is impacting their lives:

> *ABC News* wants to hear from you. We're currently producing a report on the increasing changes in our physical environment, and are looking for interesting examples of people coping with the differences in their daily lives. Has your life been directly affected by global warming?
>
> We want to hear your stories. Have you seen changes in your own backyard or hometown? The differences can be large or small—altered blooming schedules, unusual animals that have arrived in your community, higher water levels encroaching on your property.
>
> Please fill out the form below. We hope to hear from you. Thank you.[9]

The following must be what ABC had in mind. On August 30, 2006, *ABC News* reporter Bill Blakemore injected absurd editorialism in his supposed reporting, "Schwarzenator vs. Bush," claiming that after extensive research, *ABC News* was simply unable to find any debate on whether global warming is Manmade or caused by natural variability. Maybe they never went to the great academic length of putting out a website call for contrarians to match the appeal for anecdotal misfortunes, or simply "wasn't this hot when *I* was a kid" tales they expressly wanted to blame on "global warming."

As good fortune would have it, that very same day *Boston Globe* columnist Alex Beam wrote of his encounter with the alarmists' demand that their foot soldiers hew to the line of insipidly claiming ignorance, as manifested by ABC:

More curious are our own taboos on the subject of global warming. I sat in a roomful of journalists 10 years ago while Stanford climatologist Stephen Schneider lectured us on a big problem in our profession: soliciting opposing points of view. In the debate over climate change, Schneider said, there simply was no legitimate opposing view to the scientific consensus that Manmade carbon emissions drive global warming. To suggest or report otherwise, he said, was irresponsible.

Indeed. I attended a week's worth of lectures on global warming at the Chautauqua Institution last month. Al Gore delivered the kickoff lecture, and, 10 years later, he reiterated Schneider's directive. There is no science on the other side, Gore inveighed, more than once. Again, the same message: If you hear tales of doubt, ignore them. They are simply untrue.[10]

> "In one of the most remarkable signs yet of the advance of global warming, Britain's first olive grove has been planted in Devon."
>
> News item, **The Independent**, June 26, 2006

> "Over the last decades plant breeders have been developing olive varieties which grow and fruit in more temperate climes. New Zealand has been growing olives from one end of the country to the other."
>
> **Owen McShane,** director, Centre for Resource Management Studies, New Zealand

Always striving to race themselves to the bottom, ABC persisted the following month, after Senate Environment Committee chairman James Inhofe (R-OK) went to the Senate floor to deliver a speech castigating such media absurdities.[11] As Inhofe noted, "the very next day after I spoke on the floor, ABC News's Bill Blakemore on *Good Morning America* prominently featured James Hansen touting future scary climate scenarios that could/might/possibly happen. ABC's 'modest' title for the segment was 'Will the Earth Become Too Hot? Are Our Children in Danger?'"[12]

What a silly question. ABC needs to watch the news more often.

Now, despite—or, given the argument's weaknesses, quite possibly *because of*—this incessant and escalating campaign, popular belief in UFOs, astrology, ESP, and the Loch Ness Monster poll on a par with "global warming."[13] Yet remarkably the media do not flagellate themselves over the fact that heretics remain on the matter of *ET*, Nessie, and David Copperfield. Coincidentally, these phenomena offer no massive governmental intervention abetting long-standing left-wing dogma. Asteroids pose a threat more easily estimated, more readily addressed over the same century-length time scale and with a more certain risk potential, yet have produced no analogous political turmoil, with no nation being accused of failure to act. As with Nessie, no remedy for asteroids serves any interest group's agenda.

Meanwhile, the press embarrassingly fawns over a movie about Al Gore giving a slide show, while his media-darling science guru James Hansen emerges as the most oft-quoted man ever "muzzled by the administration." Ratings-starved cable news and entertainment industries pile on, with even FOX News Channel and HBO presenting alarmist specials. Unsuspecting weather junkies catch the foul odor of hyperventilation from a Weather Channel now invested in attracting alarmist-driven ratings, with their prime time "Storm Stories." The Weather Channel also brought on board a climatologist/activist[14] then, in the summer of 2006, pushed a pilot called "The Climate Code" and the transparent "It Could Happen Tomorrow." The Discovery Channel and their "Discovery-Times" joint venture with the *New York Times* piled on with an "Addicted to Oil" piece which, unable to help itself, lapses into global warming hype. Not satisfied with just one special, they tossed in "Tom Brokaw's Global Warming: What you need to know."

The latter merely proved what we *already* know: the elites are trying to out-caricature each other on environmental issues. Brokaw offered a deluge of gloom-and-doom leading the viewer not just to believe that the science is settled—remember, *if you hear tales of doubt, ignore them, they are simply untrue*—and featuring two standard global warming prophets of doom, former John Kerry campaigners James Hansen (the afore-mentioned Gore guru) and Michael Oppenheimer. Oppenheimer, having worked for years to frighten people through his pressure group Environmental Defense, informed viewers that the only scientists who don't

At the Movies

Bambi's Mother: Come on out Bambi, come on. It's safe now. We don't have to hide any longer.
Bambi: What happened mother? Why did we all run?
Bambi's Mother: Man was in the forest.

buy the alarmist line are those who benefit financially. Oddly, he didn't mention Hansen's receipt of a $250,000 grant from the charitable foundation headed by Kerry's wife, Teresa Heinz, before endorsing Kerry for president.[15]

Though media outlets find the facts inconvenient, they are not only timely but arresting. Solar activity is demonstrably cyclical, like storms. Indeed, the established forty- to fifty-year Atlantic hurricane cycle has returned as long predicted (the affected area encompasses New York City and Washington, D.C., ensuring hype for many future Storms of the Century). The "hottest decade on record" actually coincides with the closing of fully a third of the world's surface-temperature measuring stations. The glacier featured in *Time* magazine as an alarming example of melting sits mere miles from one that is advancing, if ever so lonely for lack of interested media.

Finally, to remove any doubt about the media's addiction to alarmism, facts-be-damned, consider the cover story of the UK-based *Economist* magazine, "The Heat Is On," on September 9, 2006, which actually bit on a Greenpeace scam hook, line, and sinker exposed just a few years back by the *London Mail on Sunday*[16] under the headline: "Scientists dismiss Greenpeace pictures as stunt—Global warming claim meaningless as glacier photos show 'natural changes in shape.'" The *Economist* however

"Mostly cooler and wetter than normal this July in Alaska"

"The summer of 2006 is shaping up to be quite a change from the previous two summers in Alaska with generally cooler and wetter than normal weather. Temperatures were below normal for much of the state this July, with the exception of the southern interior and parts of the southeast panhandle. The locations with cooler weather also had more rainfall than normal, and where it was warm, it was also dry. This is a good example of the typical inverse relationship between temperature and precipitation during the summer in Alaska."

Alaska Monthly Summary, July 2006

fell for Greenpeace's pitch, wistfully posting in its special alarmist pull-out two pictures of a Svalbard glacier called Blomstrandbreen, from 1918 and 2002.

Unfortunately for those who care about such things, Danish professor Ole Humlum from Oslo University, who used to work at the university's branch in Svalbard, had long ago revealed the photos as an alarmist hoax. "Blomstrandbreen is a so-called galloping glacier, which periodically advances and retreats, regardless of the climate."[17] This duping of an eager *Economist* hungry for evidence, of any sort, apparently, confirming their thesis tells us nothing, of course, except that a claim fitting the agenda is too good to check out. Research is so passé. That's what pressure group press releases are for, silly.

Historical hysteria

Today's media alarmism is nothing new. In their publication "Fire and Ice: Journalists have warned of climate change for 100 years, but can't decide weather we face an ice age or warming," the Business and Media Institute assembled a devastating array of media inconsistency in their alarmist claims, never with apology, explanation, or even recognition of their reversal(s) or even the fact their alarm has fairly accurately tracked temperature up and down.[18] Most stunning is the fact that the very reversal of Man's fate—and back again—has somehow avoided coverage.

BMI revealed that the *New York Times* has engaged in at least four separate campaigns about climate change, depending on how one views them, since warning of a new ice age in 1895. Though coverage dropped off, that warning was jump-started anew in 1924, and reversed again a decade later in 1933 with an article declaring, "Temperature Line Records a 25-year Rise." Possibly embarrassed by this gaffe, the *Times* swung back to anxiety over cooling in 1975 ("A Major Cooling Widely Considered to Be Inevitable"), and returned as is well known to its current obsession with Manmade global warming.

A humorous footnote to the *Times*'s coverage is that in April 2006, in the online edition of a Revkin story which the paper brazenly titled: "Yelling Fire on a Hot Planet: Global warming has the feel of breaking news these days," the *Times* raised the issue of its past coverage of "global warming"—emphasis on "warming"—stating that "Global warming has been a concern for several generations. Here, newspaper articles from 1956, above top, and 1932."[19] The *Times* made no mention of its "cooling" advocacy or otherwise its embarrassing promotion of alarmism over the years. You know it's bad when they even airbrush their own contribution from the record.

Time magazine initiated its own ice-age breathlessness in 1923, sensibly attributing its panic to rational factors including the sun. It then

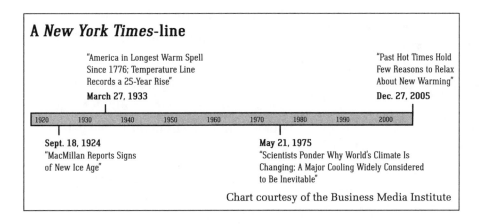

A *New York Times*-line

"America in Longest Warm Spell Since 1776; Temperature Line Records a 25-Year Rise"
March 27, 1933

"Past Hot Times Hold Few Reasons to Relax About New Warming"
Dec. 27, 2005

1920   1930   1940   1950   1960   1970   1980   1990   2000

Sept. 18, 1924
"MacMillan Reports Signs of New Ice Age"

May 21, 1975
"Scientists Ponder Why World's Climate Is Changing; A Major Cooling Widely Considered to Be Inevitable"

Chart courtesy of the Business Media Institute

reversed its position in 1939 to "weather men have no doubt that the world at least for the time being is growing warmer." Then, as now, no doubt at all. And weather men wonder how they got such a bad name.

Time then flipped back to the ice age coverage; presumably aware of the misery that cold periods have historically wrought and the boon that warming generally proved, its reportage was more panicked than its warming stories. That wouldn't last.

In 2001, *Time* popped its "frying egg" cover story about Man devastatingly cooking the planet, joining the media fury over Bush expressing agreement with Clinton's refusal to seek approval of the Kyoto Protocol. This was a warm-up for the "Be Worried. Be Very Worried" cover trotted out to help (or ride) Gore's movie hype of 2006.

The reliably alarmist BBC spent early 2006 pushing "Climate Chaos Season"[20] and enlisting its "Blue Peter" children's show to scare the lads and lassies.[21] The broadcaster's website declared:

> For the first time in its history BBC's Blue Peter is changing its name—after 4,180 episodes and 47 years Blue Peter will go "Green" on Wednesday 24 May.
>
> For one day only Blue Peter will become Green Peter to look at the changes that are happening to the planet and look at ways the Green Peter audience can help limit the effects of climate change.
>
> Forming part of the BBC's Climate Chaos season the show—on BBC ONE at 5.00pm—will come from the Green Peter garden and will look at a variety of subjects including alternative energy.
>
> The team will even attempt to boil a kettle with a bike.

At the Movies

The Future:

The polar ice caps have melted, covering the Earth with water. Those who survived have adapted to a new world.

✳✳✳

"The ancients—They did something terrible, didn't they? To cause all this water. Hundreds....Hundreds of years ago."

From *Waterworld* (1995).

In 1974 the BBC gave us "The Weather Machine," a global cooling special co-produced with the U.S. Corporation for Public Broadcasting.

News events have long spurred media interest in declaring climate alarm: even the Titanic sinking from hitting an iceberg prompted an October 7, 1912, front-page warning in the *New York Times*, "Prof. Schmidt Warns us of an Encroaching Ice Age."

Poor Professor Schmidt, he has no idea the fame, fortune, and stature that awaited him at Stanford University if he could have held on a little bit longer.

In sum, media alarmism, like climate itself, runs in cycles:

Catastrophic Manmade global warming:	1981–present
Catastrophic Manmade global cooling:	1954–1976
Catastrophic global warming:	1929–1969
Catastrophic global cooling:	1895–1932

The actual cooling and warming of our climate typically precedes the alarmism about it, and often the media are still running in the hot direction while the climate is already turning toward cold. At other times, while the media and the academy scramble to get into lockstep, we see a gap between periods of alarmism. Possibly, in this information age, some publications might actually be a little squeamish about such aggressively craven abuse of the previously noble pursuit of journalism, if not so squeamish as to miss the boat.

Gregg Easterbrook took one of the most interesting stances on this issue in the early 1990s. Whether the Earth got hotter *or* colder, we could blame it on Manmade global warming. In a piece titled "Return of the Glaciers" Easterbrook explained that there was consensus on this issue:

> The advent of a new ice age, scientists say, appears to be guaranteed. The devastation will be astonishing. Many of the world's great cities will be crushed to rubble; most of the world's agricultural breadbasket will become wind-swept

tundra; countless species will fall extinct as their habitats are frozen out of existence.

Easterbrook then proceeded to spend a majority of the article discussing the theory that global warming would actually cause the new ice age by increasing the amount of snowfall in areas where it is now too cold to snow. Greater snow cover would reflect more sunlight and more heat, prevent absorption, and thus cool the planet. He's got his bases covered at least.

Ensuring that no doubt remains about our media friends, *Grist* magazine asked "top environmental journalists," and apparently some others as well, about what problems they confront as journalists given the temptation to become advocates, as "climate change and other problems intensify."[22] The answers ranged from typical to pathological.

Consider the confused Elizabeth Kolbert of *New Yorker* magazine, who said: "I don't think the problem is with the normal standards of journalism, I think the problem is with how those standards have been interpreted. Every reporter who covers climate change knows—and has known for years—that the handful of so-called scientists who reliably try to cast doubt on the seriousness of the problem are not credible. The way I understand things, journalistic balance does not require giving equal time....Why reporters have continued to quote [skeptics] *as if* they had a claim to scientific objectivity, I'm not sure. But I don't think the corrective is for reporters to become activists (though I'm certainly in favor of

"Since the early 1990s, the columns of many leading newspapers and magazines, worldwide, have carried an increasing stream of alarmist letters and articles on hypothetical, human-caused climate change. Each such alarmist article is larded with words such as 'if,' 'might,' 'could,' 'probably,' 'perhaps,' 'expected,' 'projected,' or 'modeled'—and many involve such deep dreaming, or ignorance of scientific facts and principles, that they are akin to nonsense."

Bob Carter, "There IS a problem with global warming—it stopped in 1998," *The Telegraph* (UK), April 9, 2006

their doing so if they want to). It is simply to apply the standards of journalism—fairly and rigorously." (emphasis inexplicably in the original).

If you've gotten to this point in the book, no further discussion is required. This Kolbert Report speaks for itself: the more this so-called environmental journalist speaks or writes, the more she does indeed raise legitimate questions about who is qualified to comment on this topic.

It came from Hollywood

What child was not traumatized by the *Bambi* message of specie-ism, and horrid humanoids defiling nature and its true inhabitants? Recall the movie's rainstorm set to music suspiciously followed by two leaves dramatically dying...doubtless a subtle and premature poke at acid rain? While *Bambi* may have been the first Hollywood production of Man-as-agent-of-doom in the environmentalist mold, it has proved to be far from the last. Granted, Felix Salten's original text was as cuddly of humans as Al Gore's *Earth in the Balance*, but at least *Bambi* had some basis in reality.

Years later, *Soylent Green* brought us a world in which the "greenhouse effect" had made the planet nearly unlivable, with resulting mass depopulation despite the reality that must have been obvious even then that the carbon-dioxide fed "greenhouse effect" results in greater crop yields.

The 1970s marched on while environmentalists dithered between the storylines of Man-brings-Ice-Age and Man-fries-like-egg. Hollywood

> "The 'Green Century' special Earth Summit edition of *Time* magazine named [enviro activist Vandana] Shiva one of the environmental 'heroes,' and lauded her for representing 'tradition's voice,'" which, according to anti-poverty activist Barun Mitra awarded Shiva's promotion of "[going] back to those days when people died like flies because traditional agriculture could never feed the Indian population'" and Indian famines were hostage to drought.
>
> "Green Activist Accused of Promoting Famine Wins *Time* Magazine Honor," CNSNews.com, September 17, 2002

labored to ensure that the public did not become complacent, blessing us with *The China Syndrome*, a tale of a dogged reporter determined to out the truth about an accident at the local nuclear plant, a mishap no doubt endemic among operators combining capitalist greed with Dr. Frankenstein's madness and conceit. *The China Syndrome* had the luck of the blarney in that, mere days after its release, something forever known as "Three Mile Island" occurred. That was back in the day, when one needed luck on their side for such helpful media promotion. Now, no matter what happens—heat, cold, rain, drought, snow, storms, Hollywood can always blame it on "global warming"—such hype is *de rigueur*.

1995 brought us *Waterworld*, which was the most expensive movie ever produced prior to *Titanic*. Like *Titanic* the ship, *Waterworld* sank—and quickly—but not before getting in its green licks via the plot of a greenhouse-flooded planet and Dennis Hopper's character referring to a picture on the wall as "Old Saint Joe"—actually Joseph Hazelwood, infamous captain of the oil tanker ExxonValdez, which itself has a cameo in the flick. Even Kevin Costner, madly piloting a souped-up catamaran in pursuit of a mythical Dryland tattooed on a girl's back, couldn't save *Waterworld*.

Later in the '90s, someone named Rock Brynner, apparently the son of Yul and a teacher at Columbia University, made waves in the industry rags for selling a screenplay featuring the usual dog's breakfast of overpopulation and global warming alarm, a screenplay predicted "to do for global warming what *China Syndrome* did for nuclear power" (halt its growth for decades?). I am confident the film was never made, though awful book reviews for Brynner's *The Doomsday Report* are still cached on Amazon.com.

Testing the waters, so to speak, to see if the public had been sufficiently softened up to accept environmental pessimism as plausible plot material, a Steven Spielberg apparently running low on new ideas in the titling department followed *ET: The Extra Terrestrial* with *AI:*

Artificial Intelligence. The *AI* team got the memo to lecture us on wretchedly excessive suburban American lifestyles putting at risk the responsible clubbing, jet-setting, and of course hand-wringing of our entertainment elites.

How about this for sounding familiar, from *AI's* promo?

> Those were the years when the icecaps melted due to the greenhouse gases and the oceans had risen and drowned so many cities along all the shorelines of the world. Amsterdam, Venice, New York forever lost. Millions of people were displaced. Climate became [sic] chaotic. Hundreds of millions of people starved in poorer countries. Elsewhere a high degree of prosperity survived when most governments in the developed world introduced legal sanctions to license pregnancies. Which was why robots, who were never hungry and did not consume resources beyond those of their first manufacture were so essential an economic link in the chain mail society.

Much of this sounds like a rabid environmentalist's wish-book. Regardless, millions came to see this yarn of "A highly advanced robotic boy

Lefty obsession with "root causes" extends beyond social twaddle

"It [the ice age already under way] is the root cause of a lot of that unpleasant weather around the world and they warn that it carries the potential for human disasters of unprecedented magnitude."

Fortune magazine, 1974, writing about purported global *cooling*, such claptrap which actually won *Fortune* a "Science Writing Award" from the Institute of Physics.

long[ing] to become 'real' so that he can regain the love of his human mother." All of this sounds more like Hollywood desperately wanting Manmade global warming to be real.

Speaking of a dearth of creativity in titling these pooches, along came the most wonderfully if unintentionally humorous example of template celluloid hysteria, *The Day After Tomorrow*. Remember that early-'80s nuclear scare-fest, *The Day After*? Get it? The terminal pessimism of the nuclear freeze movement came to life, or rather death, in a scenario of a Manmade nuclear winter. Add a word to the title, sort of a *yeah, it's been a while since we trotted this one out* nod, and replace nuclear winter with carbon-dioxide winter. As one listserve colleague of mine noted, "the two are practically the same movie, and cry out to be released as a boxed set (maybe the *Time-Life Catastrophe Series*?)"

This vehicle pushed the bounds of "edgy" in that viewers were supposed to believe that Man's cruel hand at the helm of an SUV had forced hurricanes to finally repeal the remaining laws of physics and now form over land. Worse, Dennis Quaid portrayed a climatologist in the mold of the ritual Robert Ludlum academic who you just don't want to push, for he can be a physical dynamo. Quaid eschews murder in favor of deadly earnestness, trekking on foot up the arctic route of I-95, from Washington to New York. His objective: rescue his son from permanently wintering in Manhattan, a condition Quaid predicted at a Kyoto-style confab which a Cheney look-alike U.S. vice president dropped in on to be the caricature of the evil climate skeptic.

Al Gore is only the current model for the Hollywood Left's obsession with blaming you for ruining their serene slumber (made peaceful in the

---

At the Movies

"Global warming is a calamity, the effects of which will be second only to nuclear war."

Annette Bening's grating greenie lobbyist in *The American President* (1995), Hollywood's wet kiss to their idealized Bill Clinton *sans* matrimonial baggage.

knowledge that extreme poverty and primitive lifestyles exist somewhere, within one fill-up on a Gulfstream G550). His movie, *An Inconvenient Truth*, the subject of Chapter 10, is the tale of a man and his
oh-so-hip Macintosh laptop braving the lion's den of college campuses to
preach green gloom. Now, *that's* a Profile in Courage.

Chapter Nine

✳ ✳ ✳ ✳ ✳ ✳ ✳

THE BIG MONEY OF
CLIMATE ALARMISM
ENVIRONMENTALISM FOR PROFIT

S mack in the middle of the second coldest January ever measured in New York City, on a day with a low of 2 degrees Fahrenheit, Al Gore explained to a huddled MoveOn.org crowd why doubt persists about his promised catastrophic Manmade global warming:

> Wealthy right-wing ideologues have joined with the most cynical and irresponsible companies in the oil, coal and mining industries to contribute large sums of money to finance pseudo-scientific front groups that specialize in sowing confusion in the public's mind about global warming.

Gore may have put it more concisely than most, but that is the standard accusation filed against those who would question the math behind Gore and Jacques Chirac's quest toward global governance and energy rationing. Those who follow the slogan of England's Royal Society—*Nullius in Verba*—and refuse to take Gore's words on authority can soon expect to be branded as lying shills for big business.

The battle lines are clear for the media and the politicians. On one side you have brave and altruistic greens operating solely out of a love for the planet and for science. On the other you have cynical hacks working for profit.

Guess what?

✲ Enron lobbied Gore and worked with green groups in support of the Kyoto Protocol on climate change.

✲ Those "responsible" Big Businesses happen to stand to profit from otherwise ineffective environmental laws.

✲ Green groups get millions from major corporations.

To keep alive this myth, the greens need an explanation for those businesses which push the green agenda. *Why, they're just being responsible*, they tell us—as if by lobbying for new rules, subsidies, and mandates they are actually sacrificing profits. Piercing this green myth requires very little digging.

Enron: Leader of the Axis of E's

The Kyoto Protocol, that icon of green selflessness entered into by the more responsible among our world's politicians who managed to rise above corporate greed, is "precisely what [Enron was] lobbying for" and "will do more to promote Enron's business" than just about any of the numerous regulatory schemes they were pressing for in Washington. That at least, was the conclusion of an internal December 12, 1997, memo from Enron's Kyoto emissary John Palmisano, who had just returned from the completed negotiations in Kyoto where Al Gore airdropped in to impose "increased negotiating flexibility" on the U.S. team. Palmisano breathlessly closed his missive with "This treaty will be good for Enron stock!!"

Just what they wanted and good for their stock, maybe, but apparently not good enough. Enron, like many of the very biggest businesses in America, saw Kyoto—and still see global warming laws and regulations—as the best price-fixing and subsidy-creating deal in history.

Indeed, one would never know it after the scrubbing of bits and bytes, but Enron was the marquee member of the renowned Pew Center on Global Climate Change's Business Environmental Leadership Council (which membership is the quickest way to brand your company as "responsible"), Ken Lay was the poster child for the Heinz Center for Science, Economics and the Environment, and they all worked together on Enron lobbying priorities.

During my brief stint in Enron's Washington, D.C., office in 1997, I was instructed to advance their policy priority of ensuring an international treaty capping carbon dioxide. I met, or stumbled into others' meetings, to plot with green groups. I raised questions immediately—which were not well received—and departed soon thereafter. So began my education of the global warming lobby as the *Axis of E's: Environmentalists, Europeans, and Enron.* Enron has since departed, though their rent-seeking shoes have been amply filled. Strangely, however, Enron's scheming to artificially restrict energy sources through government rationing of emissions found almost no place in the extensive media coverage of its manipulative business practices.

The media were right about Enron's ethics, but they missed the best examples of the company's unscrupulousness. It is quite possible that the most emblematic among them were Enron's Kyoto games. At all levels the company would lobby and connive with green groups and like-minded Big Business to put power-hungry America on an energy diet through the Kyoto Protocol or legislation to the same effect. Meanwhile the company steadily bought up businesses to provide those renewable energy sources that Kyoto would force-feed on the population of the developed world. The Pew Center bragged about its favorite child, "Enron Wind Corp. is one of the world's largest operators of wind-power generation" (assets now owned by GE). Enron also half-owned the world's largest solar energy venture (with Amoco, now BP).

Don't forget the billions to be made when demand soars for Enron's gas pipelines—a network that at the time was second in size only to Russia's Gazprom—once coal was regulated out of business (talk about conspiring to drive up prices on consumers). Add to this the emissions-credit trading scheme Enron planned to exploit for billions more—just like the scheme presently siphoning off hundreds of millions of dollars from Europe's energy consumers with no environmental benefit—building on their success playing bookie to an earlier cap-and-trade scheme in emission credits of sulfur dioxide (an actual pollutant). Top it off with Enron's building (with U.S. subsidies, of course) coal-fired power plants in poor countries not covered by Kyoto's restrictions—plants that would no longer have to compete so much with Europe for their coal.

Add it up and that's real money.

Thanks to columnist Robert Novak, these truths did emerge early in the aftermath of Enron's downfall.[1] When Enron's green side was exposed, and the stain of profit-seeking was now on the Kyotophiles' side, Salon.com writer Timothy Noah made the sensible if apologetic point: "the mere fact that Enron stood to benefit financially from the Kyoto Treaty, and therefore was pushing energetically for its passage, doesn't in itself constitute an argument against the Kyoto Treaty."

In other words, Noah contends that arguments and policies should be evaluated on their merit, not merely on who stands to benefit financially. This was perhaps the first time an environmentalist ever uttered this notion.

The revelations from internal Enron memoranda and other sources, though still largely ignored in the policy debates, remain breathtaking even to the fairly cynical. Much of this information is largely accessible in the public domain, but snake-oil salesmen like Al Gore and his green friends managed to spout pieties about George W. Bush's energy and environment policies being the product of Enron's influence. On Earth Day 2002, Gore explicitly accused Bush of pursuing "Enron's agenda." In truth, Bush opted *against* Enron's energy-rationing scheme.

Consider what internal memos revealed or reminded us about Enron's lobbying efforts and insider ties not with Bushies but Clintonites, and the Clinton administration's cooperation:[2]

* An August 4, 1997, Oval Office meeting with *President Clinton and Vice President Gore* including Enron's Ken Lay and BP's John Browne to develop the administration position for the upcoming negotiations in Kyoto.
* A July 1997 White House meeting with Enron and other cozy industry buddies, again *including both Clinton and*

Gore, regarding the shared administration/Enron case for policy action on the theory of severe Manmade climate change.

❃ A 1997–1998 outreach campaign by the Clinton administration, employing cabinet officials to recruit further "responsible" industry and detailing the fortunes to be made from instituting the theory of Manmade climate change as government policy.

❃ In the first months of his presidency President Clinton named Lay to his exclusive panel for insiders, the Council on Sustainable Development, which would have influence on the government's relevant policies in which Enron was so heavily vested.

❃ A February 20, 1998, meeting between Ken Lay and Energy Secretary Federico Peña, staffed by the Special Assistant to Deputy Secretary Betsy Moler and Peña's chief of staff, addressing Enron's lobbying/policy desires regarding the Clinton administration's approach to restructuring the electricity system, specifically legislative positions and strategies and whether to include "climate change" policies in any such effort.

❃ Private meetings, in the run-up to the Clinton administration finalizing its Kyoto negotiating position, between Enron employees John Palmisano and/or Mark Schroeder, and senior administration officials at the Department of Energy, State Department, and EPA.

❃ Also, internal documents from the "Clean Power Group," including Enron, El Paso, Calpine, NiSource, PG&E National Energy Group, and Trigen Energy reflect they coordinated with Environmental Defense, Natural Resources Defense Council, Clean Air Task Force, Sierra

Club, and the following industry trade groups among others: Interstate Natural Gas Association, Gas Turbine Association, Solar Energy Industry Association, American Wind Energy Association, American Gas Association, and Business Council for Sustainable Energy. This coalition sought policies implementing Kyoto-style energy-use limitations without first obtaining Senate ratification of the treaty.

According to *Time* magazine, Lay was an appealing partner for Teresa Heinz because "Ken Lay and Paul O'Neill (another trustee) believed in global climate change. Ken Lay was doing some interesting things in his company about alternative energy policies."[3] Yes, one could say that.

Other correspondence surfaced, including embarrassing pleas from Heinz's assistant to Lay that he lend his image as an icon of responsibility to their global warming endeavor. O'Neill was also cited in related correspondence among high-profile Heinz Center cheerleaders, even as having promised to be this crowd's Man-in-the-Cabinet, by former Clinton State Department official Tim Wirth—famous for saying *"We've got to ride the global warming issue. Even if the theory of global warming is wrong, we will be doing the right thing."* O'Neill was ushered out after an undistinguished tenure as a dawdling treasury secretary strangely focused on matters such as convincing his cabinet-mates about "global warming."[4]

All of this is by way of illustrating that the debate is *not* greens vs. business, but quite often greens and business vs. consumers and the economy. A common refrain in environmental discussions is *"even XYZ big business is 'responsible' on this issue."* When you hear this, do the reporter's work and ask the question that in this context seems to be so uncomfortable: *Cui bono? For whose advantage? How, at whose expense, and what has this to do with being "responsible"?* Usually you'll find the *"even"* is mere puffery to confuse an otherwise transparent example of rent-seeking.

As mentioned earlier, reporters generally have no interest in asking such questions, which simply do not fit the story template. One *Washington Post* reporter even responded to me that, *well, huh, maybe Enron wasn't all that bad, after all* when informed of Enron's Kyoto shenanigans. Whether this says more about the state of journalism or business today I leave to you.

"Beyond Petroleum"? How about getting Beyond BP?

If the beef industry ran ads arguing that poultry and pork are bad for you, some level of skepticism would be warranted. Somehow, when solar panel makers and windmill companies cry that *other* energy sources will destroy the world by bringing about climate catastrophe, they are called "responsible."

Consider Big Energy—a catch-all tag for electric utilities as well as even oil companies, both of which occasionally also seek their fortune in windmill and solar mandates. Surely, the media conclude, if one of these titans can "break ranks" we have discovered virtue.

BP, for example, claims to be no longer British Petroleum, but *Beyond Petroleum*, the notable exception among Big Energy still aggressively pursuing these "renewable" boondoggles long abandoned by others for the simple reason that they are unprofitable and cannot survive without mandates and subsidies. So, is it windmills pouring out of those corroded BP pipelines in Alaska, and a solar-panel factory exploding and killing workers at the BP plant in Texas City? Of course not; BP is no more "beyond petroleum" than Michael Moore is beyond fast food.

Understand that BP leader Lord John Browne is the intellectuals' hero for his complicity in the energy-rationing wars, and elite commentary makes clear he must be exonerated from BP's string of safety mishaps. One commentator even went so far as to intimate that the problem isn't at all that BP got lost among its green scheming and pandering, it's those darn workers that just can't execute Browne's superior vision.[5]

The BP PR has worked.[6] One Harvard Business School professor wrote in the pages of the *Financial Times*: "not even the mighty Exxon-Mobil with its army of hired-gun lawyers and lobbyists unilaterally achieve everything it needs to maximise shareholder value—not least the goodwill of a justifiably skeptical public. By contrast, John Browne, BP's quiet leader, has embraced the company's responsibilities to address global warming and invest in alternative energy sources 'beyond petroleum.'"[7]

Ah yes, corporate social responsibility, which now means either (a) surrendering shareholder wealth to anti-capitalist special interests or (b) making money by investing in inefficient or even worthless products and then lobbying for them be made mandatory. BP apparently has been too busy being *responsible* to keep its pipelines in order and plants from blowing up.

First, contrary to the claim of the fawning academic, there is nothing *quiet* about BP's ceaseless and utterly disingenuous breast-beating in projecting the image of green. In Washington, D.C., it is difficult to park one's car in a garage, read a political magazine, or watch popular television shows without an endless parade of BP trumpeting its supposed environmental superiority.

As one friend in the industry put it, "they just happen to be stuck with investments we got out of decades ago." In other words, BP is performing a sophisticated beg for a bailout, and couching it in altruistic terms.

Despite decades of promises by the lobbyists for the wind industry, solar industry, ethanol, and so on that in mere decades they will be cost-competitive and won't require subsidies and mandates, these renewable sources remain, at best, niche technologies, luxurious supplements to fossil (read "real") energy sources, and completely dependent upon govern-

A Book You're Not Supposed to Read:

The Big Ripoff: How Big Business and Big Government Steal Your Money, Timothy P. Carney; New York: John Wiley and Sons, 2006.

ment for their existence. Large utilities (and GE) actually make up the bulk of the wind power lobby; the subsidies are free money underwriting their green PR (to be fair, I am informed by one of their attorneys that they also hope that this dabbling will forestall the greens' real demand: a "renewable portfolio standard"—a government-mandated percentage of power generated from wind that would basically shut down coal- and gas-fired power plants).

So in a sense it *does* require lobbying to maximize shareholder values, as our friend gazing down from Harvard's ivory tower asserts, but only when those values are sunk in technologies that are "beyond petroleum" such as corn squeezings ("ethanol") and solar panels. The real world of business generally does not work that way, as not every company can or wants to be *über*-welfare case Archer Daniels Midland, dependent not upon the marketplace but policies served up *à la carte*. Still, the media compete to bestow the aura of "responsibility" upon "capitalists" with a green hue in complete disregard of the obvious common denominator, that the companies' "green" investments would under any other scenario be painted by the same press as cynical greed rewarded with government favors born of undue influence.

Finally, it is just plain silly to profess, as our friend did in the pages of the *Financial Times*, that such rent-seekers reject the supposed bare-knuckle pressure tactics of their counterparts in favor of gentle ministrations to nudge policies toward their right and just direction. In truth, the number of in-house and outside counsel constituting BP's "army of hired-gun lawyers and lobbyists," as reported to the "Washington Representatives" listing, happen to be *double* that of the "mighty Exxon-Mobil" singled out by our professor as presenting the unseemly counterpart to the benighted BP's "soft power." Apparently the glare off the ivory can combine with the green haze of eco-posturing to obscure even the most educated of gazes.

Chemical dependence: DuPont

Delaware-based DuPont is also quite vocal in the universe of "responsible" big businesses with a large lobbying team and hefty campaign contributions to grease the skids.

DuPont regularly calls on governments across the world to do something to help reduce greenhouse gases, in the name of the looming climate crisis—and "responsibility"—often though not exclusively through its membership in the lofty Pew Center where it sat at a table with Enron.[8] Curiously, the "something" in this case (as in most other examples of rent-seeker "responsibility") is not the course of action that would more efficiently reduce GHG emissions: energy taxes, CO_2 taxes, or direct rationing of some sort. No, DuPont prefers "cap-and-trade" schemes found in Kyoto and many domestic legislative proposals.

DuPont seeks government quotas for emissions of CO_2 and other greenhouse gases from energy consumers. If a company is going to come in under its quota, it would be able to sell its remaining "emissions credits" to anybody who feared exceeding his quota.

Such a policy might not "do something" for reducing GHG emissions, but it would certainly have a discernible impact on DuPont's bottom line. And DuPont did not earn its credits as a result of an environmental agenda to reduce greenhouse gases. Instead it seems they were reducing emissions anyway, as an accidental side effect of reorganization, and knew that if they could enact a quota scheme these GHG reductions would be worth something. If DuPont wants to earn more credits, they can move their Canadian factories to any among the 155 countries exempt from Kyoto. In fact, DuPont currently has facilities in Kyoto-exempt India.

At one Kyoto negotiation, I had the pleasure of witnessing a DuPont representative—a grown man—whining before a crowded room about the U.S. refusal to sign away its energy sovereignty to the Kyotophiles.

It turns out DuPont would have potentially tens of millions of metric tons of carbon equivalent from its reductions of the covered GHG nitrous oxide as a result of a change it undertook in one of its manufacturing processes (specifically, for the chemically inclined, converting its adipic acid process).[9] As has so often happened, that tempts a company with pushing a scheme that would simply provide them *money for nothing!* What's not to love? *I did the "right thing"* . . . now *pay me for it!*

At current prices in the European Union for a ton of carbon equivalent, imposing DuPont's wishes on the American consumer would bestow a windfall of several hundred million dollars upon DuPont, with others suddenly needing to buy these "credits," all due to bureaucratic fiat and to absolutely no environmental end. Their newfound customers would be other manufacturers who wished to continue using, or increasing, the amount of energy used to conduct their business, be they energy users or consumers. Obviously, cost increases are passed on to end-use consumers just as today's baseline energy costs are passed on. Such a requirement wouldn't change much behavior, as Europe has discovered the very expensive way. But it would make money for a select few industrial mandarins.

No utility: Cinergy (now Duke Energy)

This Cincinnati-based utility recently acquired by Duke Energy may be the best example of a Republican-connected company stepping forward to fill Enron's shoes as a rent-seeker, or "responsible" big business.

Before being bought out by Duke, Cinergy's power fleet was 78 percent coal-fired[10] and quite old, requiring fairly near-term replacement. Cinergy was naturally confronted with the decision whether to replace the besieged fuel source of coal with less CO_2-intensive gas-fired capacity, but why upgrade for free when you can get paid for it? That's where a Kyoto-like plan comes in.

The switch to gas would leave Cinergy flush with GHG "credits" in the event the U.S. put itself on a carbon diet. Coincidentally, Cinergy for years has demanded a carbon dioxide "cap-and-trade" scheme.

More boldly but for different reasons, Cinergy's acquiring company, Duke Energy, became the first in its industry to call for a carbon tax. Much better than Cinergy's proposed complex scheme, Duke simply demands what in the final analysis is a competitive advantage be built into the law for Duke's formidable portfolio of nuclear and gas generation. But what about those high-CO_2 coal-fired plants that are still online for a while? Many of them are in regulated (read: government-enforced monopoly) energy markets where customers have no choice but to pay Cinergy's prices—even if they rise with a CO_2 tax.

Cinergy's corporate stance on global warming is so dignified that, at a November 2005 taxpayer-financed "workshop" to address the science of global warming (why should it take two days just to say it is "settled"?),[11] Cinergy somehow placed a senior spokesman on the panel of otherwise exclusively government-official keynote speakers, reminiscent of a parody on *The Simpsons* of Laramie tobacco sponsoring a little girls' beauty pageant. He said anyone who disagrees with Cinergy's position that Man (i.e., Cinergy) is causing global warming is a "flat-Earther," and called for a grassroots campaign to enact Cinergy's desired legislative outcome.[12]

What a corporate PR flack, one Mr. John Stowell, was doing on this dais with senior bureaucrats is anyone's guess, though the bureaucrat in charge, a Dr. James Mahoney, was visibly uncomfortable at the polemics and took to the stand, seemingly in apology for this embarrassment, to inform the audience of a "well-provisioned" open bar and hors d'oeuvres that evening thanks to some unnamed but generous contributor.

Rarely if ever in the fawning coverage of Cinergy, or their ilk, can the press muster curiosity about whether these insights and virtues derive from the profit their advocated policies would yield.

It may be helpful to know that Cinergy had recently been sued for itself purportedly causing global warming. At the time it wisely and successfully argued the absurdity of being held responsible for what it would now appear Mr. Stowell and his company believe they are in fact causing. With a race to the bottom among Duke, Cinergy, and the rest, Enron starts to look better. Or, at least, like not so much of an aberration.

Big Business and Big Green

So while the media continue to assume that the views of the skeptics are formed by payments from big business and that compliant big business is simply being altruistic, the big money continues to be bet, not on global warming, but on global warming policies.

"BP's woes are a good example of the real world dangers of corporate executives being misled by the charms of corporate social responsibility and consequent approval by the chattering class and NGO activists? The emphasis by top executives on looking green (and perhaps even being green) sent a message throughout the company that producing oil was yesterday's business, not worth the attention of ambitious managers....

Clearly, these unfortunate events [in BP's safety record] were not intentional, but they do indicate the risks of a naïve commitment to corporate social responsibility. If BP management truly believe they are 'beyond petroleum,' they should get out of the oil business and turn it over to people who are dedicated to it.

Customers still wishing to purchase BP's passé petroleum products need not worry, however. If your local station runs out of petrol because of the shutdown in Alaska, just ask them to fill up your tank with solar power."

Letter to the Editor sent to the *Financial Times*, by **Myron Ebell**, Competitive Enterprise Institute, responding to *FT*'s August 2006 coverage of BP's foibles.

In October 2006, Morgan Stanley pledged to invest $3 billion in greenhouse gas credits.[13] The *Financial Times* reported that the investment bank took this plunge, "amid mounting evidence that some US states are growing more sympathetic to international action." Shoring up this $3 billion investment is their $2 million a year lobbying budget in Washington—and who knows how much in the state capitals.

Morgan Stanley is nakedly positioning itself to get rich off government programs and then lobbying for the government programs. Again, it's lost on the media.

In California billionaire Hollywood playboy and John Kerry supporter Steve Bing spent $41 million on Proposition 87, the failed ballot initiative that would have slapped a new gas tax on drivers and dedicated the money toward renewable fuels such as ethanol. Bing, who originally inherited his wealth, has poured a ton into moonshine—in this case ethanol. Bing's partner in pushing Prop 87, Vinod Khosla, is also long on ethanol.

General Electric launched its "Ecomagination" campaign in Washington, D.C., promising to use renewable fuels, windmills, and do other Earth-friendly things as long as the government guaranteed these green strategies would be profitable. A European website reported in October 2006: "John Krenicki, president and CEO of GE Energy, a power equipment supplier involved in wind turbines, clean coal, and other clean energy technologies, said he believes that curbing greenhouse-gas emissions is 'the right thing to do.'" When the website reported that GE is heavily invested in wind, nuclear, and other "clean" technologies, it said this was "in anticipation of the [European] carbon-trading process inspiring a similar measure in the US." Presumably this "anticipation" had nothing to do with GE's $24 million lobbying budget (more than the lobbying budget of the three largest American oil companies combined).

Duke Energy, DuPont, Morgan Stanley, GE, and Steve Bing all are trying to get rich off environmental policies—by basically investing in something worthless (CO_2 credits, ethanol, wind) and then lobbying to make it mandatory. If Al Gore gets his way, they'll strike it rich, at your expense. If only Enron had stuck around a bit longer, they could have cashed in, too.

Chapter Ten

✳ ✳ ✳ ✳ ✳ ✳

AL GORE'S INCONVENIENT RUSE*

THE MAN WHO WOULD BE KING
MEETS *WATERWORLD*

"I believe it is appropriate to have an over-representation of factual presentations on how dangerous [global warming] is, as a predicate for opening up the audience to listen to what the solutions are."

Al Gore, *Grist* magazine, May 9, 2006[1]

Like the climate, Al Gore's climate alarmism is cyclical. He has a mixed history of taking the lead in declaring a CO_2-induced apocalypse, and then fleeing from his own hysteria.

In 1988 Gore ran for president to confront global warming.[2] Then he wrote a book which, despite itself, somehow assisted his quest for the vice presidency. After eight years in that office, during which time he ensured the U.S. signature upon the Kyoto Protocol despite unanimous Senate instruction not to, he went silent on the issue as he pursued the Oval Office. Gore has since indicated with remorse that this reticence was a result of being over-handled, and that it was the spinmeisters who just

Guess what?

❈ Al Gore says global warming is melting the "Snows of Kilimanjaro," but the mountain is getting colder.

❈ Gore warns of 20-foot sea-level rises, but even the UN says sea level will continue its slow and steady rise, anywhere from four inches to under three feet in the next century.

❈ Although almost all human GHG emissions have occurred since 1930, sea level has been rising since long before then, at a rate that has not statistically changed.

---

*For this chapter, I am indebted to my CEI colleague Dr. Marlo Lewis, who not only has single-handedly exposed *An Inconvenient Truth* from top to bottom, but also serves as living proof that one can indeed overcome a Ph.D. from Harvard. I also benefited from the keen eye and universal grasp of the issues possessed by Iain Murray and CEI's director of Global Warming Programs, Myron Ebell.

Get a Room

"Are there no people today as smart and honest as Winston Churchill (or Abraham Lincoln), or are they just not electable, given today's media, campaign financing, and special interests? That brings me to the Al Gore book and movie of the same name: *An Inconvenient Truth*."

The irrepressible and omnipresent—yet somehow "muzzled"—**James Hansen** in a draft submission to the *New York Review of Books*

didn't see his brand of eco-tub-thumping as the pony to ride back to the White House. Those days, Gore leaves no doubt, are now over.

After losing, being robbed, whatever, he rediscovered his inner alarmist and released a movie, *Starring Al Gore*, all about Al Gore.

Thanks to this cinematographic venture Al Gore and global warming are now forever joined at the hip so far as the public is concerned. In honor of the movie, the *Christian Science Monitor* even coined a term, "Docuganda."[3] Gore's celluloid claims are now the anecdotal and rhetorical limbs in which the greens now perch like so many redwood-squatting Julia Butterfly Hills. It is time to take a chainsaw to them.

A better title: *One Flew Over the Cuckoo's Nest*

With Gore having "gone Hollywood," a scouring of the rankings of "top movies of all time" seemed appropriate, and revealed that indeed most of the appropriate titles for Gore's film were taken: *Pulp Fiction; The Usual Suspects; Psycho; Apocalypse Now; The Green Mile; Groundhog Day; Judgment at Nuremburg; A Man for All Seasons; Planet of the Apes; The Man Who Would Be King*...the list goes on. Clearly *not* considered, however, were *Shadow of a Doubt*; or *It's a Wonderful Life*.

This 2006 movie about a slide show, *An Inconvenient Truth*, is best described as a movie about a book (*Earth in the Balance*)—from which the author spent twenty years running away—the movie itself having immediately been turned into a book (about a movie about a book, of course). We can only hope for the movie version of this latter book. Witness the joys of recycling.

Raging bull

Reviews, rather, what professional movie reviewers wrote, about *AIT* include the embarrassing love note from an overeager Roger Ebert:

> In 39 years, I have never written these words in a movie review, but here they are: You owe it to yourself to see this film. If you do not, and you have grandchildren, you should explain to them why you decided not to.
>
> —RogerEbert.com, June 2, 2006

Ebert is not alone in his schoolboy crush on this Passion of the Gore. Other "critics" added to their enduring legacy for sagacity, sitting through this one side of a hotly contested issue to emerge convinced and proclaiming that there is only one side:

> *An Inconvenient Truth* . . . succeeds at cutting through the clutter surrounding global warming by making a clear, compelling case for how our actions are affecting the planet.
>
> —Ethan Alter, *Premiere* magazine, June 23, 2006

By "cutting through the clutter" I believe he means "ignoring overwhelming evidence to the contrary." As might be expected Ebert won Best Fawning Review:

> Am I acting as an advocate in this review? Yes, I am. I believe that to be "impartial" and "balanced" on global warming means one must take a position like Gore's. There is no other view that can be defended.

Notably, Ebert offered less a review than repetition of Gore's assertions—flatly stating as truths matters about which Ebert has no particular understanding. He did occasionally deign to preface his verities with *"Gore says"* but just as often failed to so qualify these truths he parrots. Ebert does however manage to double Gore's outrage over the fact that not everyone believes him.

Apocalypse now

Gore's was no mere movie but a forceful hour and thirty minute sermon, with more brimstone (and better air conditioning) than the American churchgoer might be accustomed to. Gore's testimony affirms the beliefs of the truly passionate, the Global Salvationists: the environmentalists and their fellow believers in the creed that development and technology grind humanity's billions, plus nature herself, under their jackboots on an unholy roll toward a coal-fired apocalypse.

It is this very prediction of disaster that explains the lovefest. The entire parade of global warming horribles is trotted out, including (enormously) rising sea levels and ravaged coastal areas, increasing tornadoes, intense heat waves. These dutiful role-players all strolled to place on cue, joining drought, wildfires, and more melting things (including an ice-free Arctic Ocean by 2050).

Gore says he is a recovering politician, but his movie showed the instincts are still there as he agonized over threats appreciated by average people: cute critters and the average folks themselves. Gore predicts mass extinctions (a million species by 2050) and of course, deep inroads into the depopulation fantasy. In the great tradition of Malthus and Ehrlich, Gore warns of 300,000 human deaths attributable to global warming in little over two more decades. Those who don't die will migrate or be swamped by migrants, according to Gore's warning of "100,000,000 refugees." That's one out of every sixty or so people on the planet fleeing their own swamped homeland to squat on someone else's lawn or Superdome. (Given that there's apparently nothing that he won't attribute to global warming, that figure seems kind of low. A little ingenuity drives the numbers up further, what with inevitable revivals of brain fever, the staggers, gout, housemaid's knee, and so on. Three hundred thou is just a first wave.)

Movies are entertainment. As noted, to maintain his academic persona, Gore makes the mistake of committing his narrative of fast-moving montages to writing, *sans* soundtrack and effects, and publishing it as a book. This novelization of *AIT* by the same title[4] is pure advocacy of the legal-brief variety, not a science report. In fact, it is of the ambulance-chaser phylum of legal briefs, counting on an incurious jury. And no opposing counsel.

The book remains true to the film in that it presents a one-sided case of evidence favorable to his Malthusian argument. It often ranges from mere conjecture to pure fantasy. Scientific method, less driven by advocacy and more by exploration into relevant arguments and countervailing theories, has no place in Gore's opus.

For example, Gore continues beating the drum of global warming causing retreat of glacial and snowpack cover of Kenya's Mt. Kilimanjaro. His corny Hemingway allusions crowd out the well-documented reality that global warming probably has nothing to do with the steady disappearance of the "snows of Kilimanjaro." In fact, the area around Kilimanjaro has been *cooling*, yet the snows (ice) have been retreating for more than 100 years—long before the invention of the SUV. Instead of retreating due to warming, the ice is vanishing from declining atmospheric moisture. The list of Gore's similar transgressions is not endless. It only seems like it.

In Gore's world, there is little theory. There is only the *known* (what he believes), and background noise (skepticism, all certainly

A Book You're Not Supposed to Read

Environmental Gore: A Constructive Response to Earth in the Balance, Pacific Research Institute (Edited by John A. Baden), 1994.

bought-and-paid-for). Gore's movie presents only evidence, largely anecdotal, favorable to his political agenda. He often presents it in misleading ways not only ignoring but occasionally editing out evidence belying his alarm, even when it conclusively puts the lie to it.

Sins of omission

Before tackling the exhaustive list of affirmative misrepresentations and outright fabrications in Gore's movie, let us examine some of those things that slipped Gore's mind. Here are some of Gore's omissions about the impact of emissions:

"Sexing It Up"

***Grist* magazine:** There's a lot of debate right now over the best way to communicate about global warming and get people motivated. Do you scare people or give them hope? What's the right mix?

Gore: I think the answer to that depends on where your audience's head is. In the United States of America, unfortunately we still live in a bubble of unreality. And the Category 5 denial is an enormous obstacle to any discussion of solutions. Nobody is interested in solutions if they don't think there's a problem. Given that starting point, I believe it is appropriate to have an over-representation of factual presentations on how dangerous it is, as a predicate for opening up the audience to listen to what the solutions are, and how hopeful it is that we are going to solve this crisis. Over time that mix will change. As the country comes to more accept the reality of the crisis, there's going to be much more receptivity to a full-blown discussion of the solutions.

Al Gore, admitting in an interview that he is both overplaying the dangers of global warming, and downplaying the sacrifices he wants us all to make.

- Viewing the big picture, the film and the book present as settled science what is in fact highly contentious—the claim that global warming has significantly increased or will significantly increase hurricane frequency and severity.[5]
- Gore speaks as if we know for sure what past weather conditions and temperatures were, and never acknowledges the significant debate over whether the commonly used "proxies" are accurate measurements of temperature.
- On the policy front, Gore misleadingly implies that if the government forces us to cut back on our energy usage we will be safer from hurricanes (See Chapter 7). This is not to say that Gore merely implies his biggest whoppers, but this stance being laughably unsupportable even according to his allies may be one reason Gore only hints at it if, like the idea that CO_2 causes warming, unmistakably.
- Similarly, he refuses to explore the costs he is asking his viewers and readers to bear through policies like Kyoto. Such candid talk might prompt scrutiny of the alarmism.
- The film treats as probable or likely what most scientists regard as highly *unlikely*: the wind currents across the Atlantic will shut down (his pet scenario, shared with *The Day After Tomorrow* and now certain CNN newsreaders).[6]
- Gore utterly ignores the real reasons for increases in various weather-related damages—more people and more wealth situated in more flood-prone or storm-prone areas.
- Gore, dwelling on dire forecasts, never confronts the actual, observed rate of warming: over the past three-plus decades since we have employed modern technology, the planet appears to have warmed 0.17° C per decade, or slightly less. Gore's scary scenarios depend on far more aggressive warming than that.

�֍ Gore ignores every model that projects warming to continue at the current rate—nearly every single one among dozens—and instead focuses on the one apocalyptic outlier. Gore stays mum about the fact that even under the alarmists' scenarios warming would almost certainly be about 1.7 degrees Celsius over a century (and despite Gore's suggestions, this says nothing at all about Man's possible influence).

✖ *An Inconvenient Truth* conveniently omits that Greenland, one star of Gore's melting-ice show, was as warm, or warmer, in the 1920s than it is today, and that it was heating up faster then.[7]

✖ The former vice president never mentions that in the early part of the current Holocene Epoch[8] the Arctic was actually *several degrees warmer* than at present—*as much as 5 degrees Celsius warmer in some places.* Again, this is well established, and therefore simply elided. Gore's obvious conclusion is that if Man (somehow) causes this sort of warming, it suddenly becomes catastrophic.

✖ Gore separately presents graphs of global temperature changes going back several hundred thousand years,[9] and a chart of atmospheric CO_2 concentrations throughout a similar period. Ne'er the twain charts shall meet . . . for very good reason. Although Gore strongly hints that CO_2 emissions caused past temperature changes, putting the two charts together shows no such relation: one is up while the other is down, one precedes the other, then vice versa for both. In fact, several scientific papers indicate that rising temperatures *caused* CO_2 changes, not the other way around.

✖ Remarkably, Gore neglects to mention that this graph suggests that the previous four inter-glacial periods were *warmer* than the interglacial period in which we are now living. After all, that is very, very inconvenient. And true.

❅ *An Inconvenient Truth* abuses the Vostok ice core samples (large cylinders of ice plucked from the Arctic and analyzed to try to gauge the temperatures and CO_2 content over past centuries). He fails to note that these cores show temperatures higher than today's in previous inter-glacial periods, despite CO_2 levels at the time being *lower* than today's. In other words, relaying the ice cores' whole story would pierce his argument that Manmade CO_2 determines the climate, i.e., that Man has assumed control over the weather. So he skips it and sticks with his predetermined, faith-based conclusion.

❅ Specifically, these ice cores suggest, again, that the temperature increases *precede* the CO_2 level increases (although both sides of that debate must in good conscience admit that the resolution from this data is not good enough to afford certainty). That is probably because the oceans, which hold about fifty times more CO_2 and about five hundred times more heat than the atmosphere, can absorb CO_2 better when they are cold. When it warms, the seas release CO_2 into the air.

❅ Gore misrepresents the way CO_2 actually contributes to the "greenhouse effect." In mathematical terms, the CO_2-greenhouse effect relation is logarithmic, not linear. That is, each molecule has less of a greenhouse impact than the molecule before it. A doubling of the amount of CO_2 in the air has the same effect as the previous doubling. In short, even global warming theory holds that Man's emissions are insufficient to have caused the one degree warming since the Little Ice Age ended.

Sins of commission: The usual suspects

Gore also engages in flat-out misrepresentation, again often combined with omission. Consider the trailer designed to draw audiences in, which notably includes the following parade of horribles:

"If you look at the ten hottest years ever measured they've all occurred in the last fourteen years and the hottest of all was 2005"; "The scientific consensus is that we are causing global warming"; images of Kenya's Mount Kilimanjaro "thirty years ago and last year; within the decade, there will be no more snows of Kilimanjaro"; "Temperature increases are taking place all over the world. And that's causing stronger storms"; "Is it possible that we should prepare against other threats besides terrorists?"; "The Arctic is experiencing faster melting"; he then offers his melting scenario, after which "sea level worldwide would go up twenty feet."

Columnist Paul Stanway called the sea-level claim "[a]mong Gore's more outrageous nose-stretchers" in the film, noting that "[a] 2005 joint statement by the science academies of the Western nations, including the U.S. National Academy of Sciences, actually estimates a worst-case scenario of thirty-five inches"[10] (and the IPCC agrees that it is apparently equally likely we would see an increase of four inches). As you will see, this is typical Gore. That whopper may be among Gore's worst, but it has plenty of company. Though many of the issues were treated separately throughout this book, below are the inconvenient truths about these supposed horrors and other key missteps larded throughout Gore's meticulous if meticulously deceptive treatment of the issue.

2005 was the hottest year: In truth, satellite temperature measurement data inform us that 2005 was not the hottest year even since measurements began being taken in response to the global cooling panic in the late 1970s. Really 1998 was, and NASA's very expensive satellite monitors indicate stable temperatures since 2001.[11]

The Ten Hottest Years!: Gore says, "If you look at the ten hottest years ever measured they've all occurred in the last fourteen years and the hottest of all was 2005." This is a riff on "the 1990s were the hottest decade in history," so we will consider them together.

This particular chestnut doesn't survive even the slightest scrutiny. First, Bob Carter, a geologist at James Cook University, Queensland, Australia, gained brief notoriety in 2006 for tweaking the greens' beaks with their own "baseline game," noting that "the official temperature records of the Climate Research Unit at the University of East Anglia [UK], [show] that for the years 1998–2005 global average temperature did not increase."[12] That's right, it cooled after 1998. Gore failed to mention this. (Were Gore still on the "global cooling" bandwagon, however, his style of argument leaves no doubt that this would be screaming from movie posters.)

Second, Gore relies upon surface temperature measurements. But pre-1990 surface temperatures cannot necessarily be compared to post-1990 surface tempeartures. Consider, once again, what Canadian economist Ross McKitrick found (graph at right).

Europe's deadly 2003 summer heat wave was Man's fault: Actually, the massive deaths resulting from the hot August in Western Europe were a result of cultural issues peculiar to France, resulting from the odious mélange of a month-long holiday,

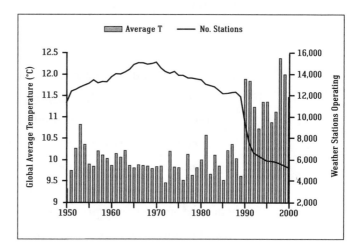

the practice of leaving the elderly at home whilst the masses (including nurses) vacate for the beach, and pricing air conditioning out of the reach of many. The heat wave was not anomalous. From the "slow-learner" files we see that during the heat wave of July 2006, the French medics were on strike.

Even the oft-reliably alarmist U.S. weather offices remain sober on this point, professorially pointing out how the 2003 heat wave itself was caused by an atmospheric pressure anomaly, not "global warming."[13] Blaming America for the shame and humiliation of horrific piles of unclaimed French dead while relatives vacationed is certainly the easier explanation.

Temperature increases are taking place all over the world: This is certainly true if by this Gore means temperature is increasing someplace, the world over. It is absurdly fraudulent if by this he intends to claim, as it appears, that temperatures are increasing "all over the world." Some areas are cooling just as some are warming, some are getting wetter as some are getting drier, and so on. That's the way climate has worked and will continue to work. Global warming alarmism relies on a claim of an increased average temperature among those places we measure. Again, however, averages don't happen, and we are always trending one way, or the other—frankly, both ways—a conflict resolved simply by one's selection of a baseline year.

Computer models relied upon by the alarmists, by the way, predict that temperatures will increase in the coldest parts of the world (the poles), in the coldest months (their winter), at the coldest time of day (nighttime) ("polar amplification," discussed throughout these pages).

No scientists disagree: This claim, precisely speaking, is a lie and a very bad one at that, as dispensed with elsewhere in these pages.

CO_2 causes warming: Except when it doesn't. As noted above, typically temperature increases preceded CO_2 increases (often by substantial periods of time); over a few periods the lines cross, such that when temperature is rising, CO_2 levels are falling or vice versa.

In 2005 testimony to Canada's Commons Committee on Environment and Sustainable Development, Carleton University paleoclimatologist Professor Tim Patterson noted that "There is no meaningful correlation between CO_2 levels and Earth's temperature over this [geologic] time frame. In fact, when CO_2 levels were over ten times higher than they are now, about 450 million years ago, the planet was in the depths of the absolute coldest period in the last half billion years."

Devastating to Gore's claim, during the movie's June 2006 opening his temperature chart appeared in an article in *Science* magazine, the abstract of which begins, "During the early Pliocene, 5 to 3 million years ago, *globally averaged temperatures were substantially higher than they are today, even though the external factors that determine climate were essentially the same.*"[14]

That is, the forces that determine temperature aren't warming us up as much now as they did when these very same natural forces were...the same. Put another way: it is cooler now than in the past with the same GHG concentrations. The global warming/cooling industry does seem to have a repetitive problem with cause-and-effect relationships. This is certainly no exception. There is a reason Gore doesn't combine the CO_2 and temperature charts for audiences.

Similarly, Gore trots out the thoroughly discredited "Hockey Stick" chart, which claims climate was stable until Industrial Man came and messed it up. He then brazenly declares the Hockey Stick has been *validated*. This is possibly one of his

Yew Can't Take It with You

"The Pacific Yew can be cut down and processed to produce a potent chemical, taxol, which offers some promise of curing certain forms of lung, breast, and ovarian cancer in patients who would otherwise quickly die. It seems an easy choice—sacrifice the tree for a human life—until one learns that three trees must be destroyed for each patient treated, that only specimens more than a hundred years old contain the potent chemical in their bark, and that there are very few of these yews remaining on earth."

Al Gore, *Earth in the Balance*, 119

FIGURE 1. FROM THE EAST, AUGUST 1911
Photo by T. W. Stanton, U. S. Geological Survey.

FIGURE 2. FROM THE EAST, AUGUST 29, 1935
The glacier consists of two separate ice bodies in contrast to the condition depicted in Figure 1.
The north moraine lies at the extreme right edge of the lower ice mass.

Grinnell Glacier

most shameless non-truths, debunked in detail elsewhere in these pages.

Frozen things have a very limited repertoire. They grow and they melt. Sometimes they retreat by growing—that is, "calving," after extending beyond their ability to stay intact—which alarmists actually cite to imply greenhouse-melting. Gore even shows video of such calving. Frozen things are performing all three acts all over the world, though you would never know that if you asked someone who wants you to drive a small car *because it's the right thing to do.*

Regarding these hunks of ice in short, if retreating glaciers are proof of global warming, then advancing glaciers are proof of global cooling. They can't both be true and, in fact, neither is.[15]

Those things that are melting have generally been doing so, and at the same pace, for over a hundred years. Al Gore has long been fond of citing the retreat in Glacier National Park as proof of the horrors of Man-made global warming. Unfortunately for him, the melting actually occurred in earnest *before* Man started adding a noticeable amount of GHG to the atmosphere.

Kilimanjaro is one of Gore's favorite props, being the romantic icon of the literati such as it is. Kilimanjaro's snow is indeed fading away. It has been disappearing for a while. What's more, the "snows of Kilimanjaro" have been receding although scholarly articles note that the area's temperature has *not* been going up. You see, you need two things for ice: cold *and* moisture. It's the latter—not the former—that is lacking.

As climate scientist Robert C. Balling points out: "Gore does not acknowledge the two major articles on the subject published in 2004 in the *International Journal of Climatology* and the *Journal of Geophysical Research* showing that modern glacier retreat on Kilimanjaro was initiated by a reduction in precipitation at the end of the nineteenth century and not by local or global warming."[16] In other words, the local climate shift began a century ago.

Revealing this truth requires time-consuming and distracting explanation, unlikely to advance Gore's Man-as-Agent-of-Doom hypothesis, and certainly not his anti-energy zeal. For example, one article makes the case that lower snowfall is being caused by deforestation. Inarguably, providing a modern lifestyle including energy supply and—gasp—pesticides dramatically reduces deforestation and, in this case, providing the locals these banes of the environmental warming alarmists might actually put a halt to Kilimanjaro's ice loss.[17]

The **Upsala** glacier, which is the subject of a pressure group scare campaign, is touted by *Time* magazine's recent "Be Worried. Be Very Worried" cover story. This mid-size Patagonian glacier and major tourist attraction is suddenly retreating, we are told, thanks to a "discovery" by Greenpeace (naturally). Upsala's behavior has been ably recorded over time by scientists at the Swiss-based World Glacier Monitoring Service, who attribute the change to dynamic causes unrelated with air temperatures.

Just fifty kilometers away from Upsala is the Moreno, which is growing, and thereby calving. Claims of Moreno's retreat are typically danced around without brazenly lying about it "melting," and are pure fabrication and illustrative of the lengths to which environmentalists go to dishonestly frighten people. As again noted by the fraud-busting *Mitos y Fraudes* project of the Fundacion Argentina de Ecologia Cientifica,[18] the area is frozen and the "retreat" that is noted is in fact "calving," which is the breaking off of a large piece of this ice river caused by *increasing* ice mass pushing the unit down to where its length and heft are unsustainable for the given topography. This occurs at Moreno every four to five years, so expect the uncritical media ruse to run at about the same frequency.

Other much larger glaciers nearby are advancing without the pleasure of angst-ridden eco-tourists and pencil-gnawing journalists. In fact, "While in the same region in Patagonia, either in the Argentinean or Chilean sides, there are small glaciers on retreat, there are other medium sized glaciers in stable condition, and really big glaciers *growing at record speed and volume.*"[19] For example, South America's largest glacier, Pio XI, is growing very quickly. The greens and media just missed it, that's all, another case of not seeing the forest for being busy hugging the trees.

Time, Gore, and their cohorts do not reveal these facts about other South American glaciers advancing, which debunk the idea that some global temperature increase is responsible[20] for their favorite glaciers melting. Given that sea-level rise is demonstrably in line with its historical behavior, it seems off target to classify a few melting things as "catastrophic." Whether we should mourn this water's passage from solid to liquid state is a matter of taste, and phobia. Remember that the greens think those very, very hot places in the world are also just perfect. Everything is perfect, that is, so long as we can't see (credibly or otherwise) any human influence on it. Is it calamity that frightens, or change?

Having dispensed with the alarm over Man purportedly melting Patagonian glaciers, let us shift over to the Andes. Al Gore tells us the Andean glaciers are melting. Unlike in Patagonia, here he is accurately representing the picture. Of course, he is misleading us about the cause.

Just as Gore's movie was hitting it big, a team of scientists led by University of Massachusetts professor P. J. Polissar published the peer-reviewed article "Solar Modulation of Little Ice Age Climate in the Tropical Andes." While the article includes a mandatory line worrying about future Manmade global warming, the actual substantive points pull the rug out from under Gore.

Published in the *Proceedings of the National Academy of Sciences*, Polissar, et al., show that man is not the culprit in the melting of the Andean glaciers. The sun is. Also, it implicitly rejects the "Hockey Stick" by agreeing there was a Little Ice Age (LIA). Finally, it admits that both warmth and glacial melting are hardly unprecedented in the Andes.[21]

The piece argues that "climactic change in the Venezuelan Andes is linked to changes in solar activity during the LIA." Later it gets more to the point, saying the data "suggest that solar variability is the primary underlying cause of the glacier fluctuations." Finally, the authors also conclude: "During most of the past 10,000 yr, glaciers were absent from all but the highest peaks in the Cordillera de Merida."

History is repeating itself. Amid the 2006 furor over the UK's Royal Society condemning those who dare disagree with their "consensus" on Manmade climate change and effort to financially isolate them, Will Alexander of South Africa's University of Pretoria piquantly reminded us of how history will likely view the modern authoritarian science. "In 1600 Giordano Bruno was burned at the stake for supporting Copernican theory of a Sun-centred universe. In 1632 Galileo was accused of heresy for supporting the Copernican view and was forced to retract. Now the Royal Society has issued an edict excommunicating all those who maintain that it is the sun and not human activity that is the cause of variations in the earth's climate."[22]

Reaffirming this revival of heliocentrism—yet while continuing its drumbeat about a horrible future Manmade climate—in September 2006 *New Scientist* magazine reminded readers of the link between "prolonged lulls in the sun's activity—the sunspots and dramatic flares that are

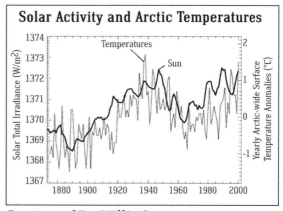

Courtesy of Dr. Willie Soon

225

driven by its powerful magnetic field,"[23] and the misery of the Little Ice Age. They noted an increasingly recognized relationship between sunspots and temperature, adding that the past fifty years of abnormally high solar activity appears headed for a slump, indicating another cooling is on the way.

Naturally, at about the same time, possibly to cast the elites' now-ritual (at minimum, when you're lucky) disapproving gaze at such inquiry, leading alarmist and Gore advisor Tom Wigley published a paper claiming that the sun has no influence.

Accepting the Wigley hypothesis that the sun has no impact on our climate requires one of three possible implications, none of which are convenient to Gore, et al.:

1) If CO_2 had such a powerful effect at low concentrations but is now having proportionately much less of an effect, as we must assume given Wigley's argument, it would be reasonable to conclude that further concentrations will have even less effect. In short, the alarmists' own arguments now corner them into arguing that Man's impact on the climate even as they posit it is waning, not increasing. Ouch.

2) If the Sun didn't drive our climate's emergence from the Little Ice Age as Wigley argues and CO_2 concentrations had the minimal effect at low levels science has always accepted, then some still-unidentified force drove this climate phenomenon. If we don't know what this is, then all bets are off on what is driving the current warming. Don't expect the greens, who *begin* their analyses with such certitudes, to select this option.

3) CO_2 is more powerful than we thought at low concentrations and so may therefore be more powerful than we thought at high concentrations, and CO_2 is therefore currently saving us

from an ice age. Again, with the beginning hypothesis—that is impermissible to challenge—being that Manmade CO_2 is the root of all climate evil, this is also not a desirable option for the alarmists.

The old adage about not lying because it makes things so complicated later, what with trying to remember what one said, reconciling the stories and the like, seems to have some utility in this instance. Claiming the sun

The Snowjob of Kilimanjaro

Al Gore: Somewhere along these last decades global warming has become not just a potentially significant problem—now it's showing up in the real world. Glaciers all over the world are melting. Within fifteen years there will be no snows of Kilimanjaro.

"Riders on the Blue Marble Must Confront Climate Change"

✳ ✳ ✳

A Real Scientist: [T]he observations and interpretations made during two periods of fieldwork (June 2001 and July 2002) strongly support the following scenario. Retreat from a maximum extent of Kilimanjaro's glaciers started shortly before Hans Meyer and Ludwig Purtscheller visited the summit for the first time in 1889 caused by an abrupt climate change to markedly drier conditions around 1880....Once started, the lateral retreat was unstoppable, maintained by solar radiation....**Positive air temperatures have not contributed to the recession process on the summit so far**....The scenario presented offers a concept that implies climatological processes other than increased air temperature govern glacier retreat on Kilimanjaro in a direct manner.

G. Kaser, et al., 2004. "Modern glacial retreat on Kilimanjaro as evidence of climate change: observations and facts," *International Journal of Climatology*, 24, 329–339.

has no influence on the climate, despite the persistence of historical records to not flee in terror at the greens' pressure campaign of authoritarian science, does leave the impression that the global warming alarmists have so contorted themselves with faith-based conclusions protected from science's harsh glare that they find themselves tied into a rhetorical and logical knot only solved through answers irreconcilable with their beginning propositions that got them in this mess.

Melting ice sheets

The supposedly melting Greenland ice sheet is what Gore says will drown us all. Although you are supposed to forget such things, during the Medieval Warm Period the Vikings grew forbs or grasses for their livestock in their Greenland settlements. That suggests Greenland was warmer then, and there was more melting. Greenland was even warmer during the climate optimum in Roman times, and we have no reason from our sketchy records of sea levels to think that there was enough melting during the latter era to raise the sea levels much.

The *western edge* of Greenland may be experiencing ice-melt—and faster the last few years, but as with Kilimanjaro we know for certain that this is *not* because temperatures on Greenland have been rising. Science reveals this to be more likely a result of cyclical changes in ocean currents. Several papers in fact reveal that temperatures on Greenland are not behaving as Al Gore would wish (or have us believe). Petr Chylek at the Los Alamos National Laboratories helped burst the Greenland-is-burning bubble in the influential *Journal of Geophysical Research Letters*, contemporaneous with Gore's movie.[24] But when some scholar points out that Greenland's inland glaciers are growing, naturally, the alarmists invoke even *that* as proof of global warming. Experience indicates the same claim would be shouted from the solar-paneled rooftops were the opposite trend discerned.

Also note that the reason the Vikings left Greenland during the Little Ice Age was not that Greenland went from being tropical to chilly (though it did indeed become cooler and no longer suitable for their agriculture). What truly impacted their flight was that the coastal areas were disappearing when the ice build-up pushed down to diminish habitable coastal areas.

Chylek addressed two recent warming periods experienced on Greenland's coast—in the 1920s and 1995–2005. The article's abstract concludes that "temperature increases in the two warming periods are of a similar magnitude, however, the rate of warming [in the '20s] was about 50 percent higher than that in 1995–2005." The text states:

> [T]o what extent can the current (1995–2005) temperature increase in Greenland *coastal regions* be interpreted as evidence of man-induced global warming? Although there has been a considerable temperature increase during the last decade (1995–2005) *a similar increase and at a faster rate occurred during the early part of the twentieth* century (1920–1930) when carbon dioxide or other greenhouse gases could not be a cause. The Greenland warming of 1920 to 1930 demonstrates that *a high concentration of carbon dioxide and other greenhouse gases is not a necessary condition* for a period of warming to arise. The observed 1995–2005 temperature increase seems to be within a natural variability of Greenland climate"[25] (emphases added).

The paper continues:

> To summarize, we find no direct evidence to support the claims that the Greenland ice sheet is melting due to increased temperature caused by increased atmospheric concentration of carbon dioxide. The rate of warming from 1995

to 2005 was in fact lower than the warming that occurred from 1920 to 1930.[26]

Oh, dear. The paper called for continued observation and pursuit of knowledge, boldly intimating that we may be years from being able to credibly assert those things which are already preached breathlessly as known absolute and indisputable truths. For daring to practice the evils of science, Chylek is welcome to borrow my food-taster and car-starter.

The Arctic ice is melting. What with the high-profile, if quietly abandoned, claims of unprecedented Arctic melting by outlets such as the *New York Times* and IPCC, Gore had to say something to keep up. He chose, "Starting in 1970, there was a precipitous drop-off in the amount and extent and thickness of the Arctic ice cap." Former University of Winnipeg climatology professor Tim Ball asserts that this represents a major sin of omission in that the claim is based on a single transect (sample) taken not only during a known cooling period but in October; then, to create the claim of melting, the sample is compared with several measurements taken later in a warmer month using completely different technology.[27] As such, the so-called comparison is actually no such thing. It is holding up one isolated set of data against another utterly unrelated sample, and claiming to derive knowledge from it where none is available. That's not science, and only a politician—recovering or otherwise—could pretend that it was.

Ball's criticism is damning if true. Despite tantrum-like insistence that "consensus" exists, as with so many areas of climate science this matter remains one of dispute and further study. The best-case scenario for Gore is that his claim of 15 percent loss of sea ice since the 1970s is correct. This would represent a major climb-down from where the alarmists were just a few years ago when the IPCC originally concluded a 40 percent loss determined from submarine data, which it then had to revise to 15 percent when it was discovered the data simply came from a transect where ice is sufficiently thin to allow submarines to crash through to the surface.[28]

Down at the other end of the planet, the ***Antarctic Peninsula*** is experiencing melting; this is the part that sticks up northward (toward South America) from the continent, extending into the Southern Temperate Zone. Again, this is a smaller, selected area isolated by alarmists as proof of something that isn't happening elsewhere. The non-peninsular Antarctic continent, by comparison, is a vastly larger area accounting for 96 percent of the Antarctic, holds nearly all the ice pack, and is *not* warming, and no research exists to claim it is melting.[29] Many measuring stations show a trend of declining temperatures.[30] The Antarctic ice sheet, the largest on the planet, might actually be *thickening*.

Regarding the peninsula, Gore falsely suggests that melting there is unprecedented and that greenhouse gases are the cause. Dr. Wibjörn Karlén, professor emeritus in the Department of Physical Geography and Quaternary Geology at Sweden's Stockholm University, admits, "Some small areas in the Antarctic Peninsula have broken up recently, just like it has done back in time. The temperature in this part of Antarctica has increased recently, probably because of a small change in the position of the low pressure systems."[31]

No climate models based on Manmade warming can explain the peninsular warming. This suggests something *natural* is at work. Recall the data indicating that the world has existed for millennia pre–*Homo sapiens*, and as such show that natural processes exist, are variable, and are more than a little bit stronger as a default culprit when assessing responsibility for that which has long occurred. On the alarmist side, however, the rationale is the anti-scientific, indeed rather pagan instinct to yell, "I can't explain this phenomenon, and so the cause must be Man."

In November of 2006, two scientists published the result of their satellite-data study and concluded that the Antarctic ice sheet was growing at 4–6 millimeters in thickness per year. Of the 72 percent of the Arctic ice sheet they studied, they estimated it was growing by about 26 million tons per year.

But whether a study finds thickening or thinning, either conclusion is cited as evidence of global warming. Both conclusions are absurd given the paucity of data measuring Antarctica's total ice mass. Gore, for example, cites a study based on only three years of Antarctic data (again, *at this pace, my son should be thirty feet tall by the age of thirty*).[32]

Sea-level rise: Too busy launching into completely fabricated hyperbole, Gore fails to remind movie-goers that sea levels have been rising since the end of the last ice age, at a rate of 1.8 mm per year for the past 8,000 years, and will continue to do so at varying rates until the next ice age. The IPCC does not forecast sea-level rises of "eighteen to twenty feet," but actually a possible range from *four inches to less than three feet* over the century, and concludes that current trends may *or may not* be slightly higher than the trend over the past 150 years.[33] For example, even the politically crafted *Summary* of the IPCC's Third Assessment is hardly ambiguous in stating "No significant acceleration in the rate of sea level rise during the twentieth century has been detected," and "Within present uncertainties, observations and models are both consistent with a lack of significant acceleration of sea level rise during the twentieth century."[34] This is despite breathless claims and photo montages purporting unprecedented melting *already*.

Meanwhile, the International Union for Quaternary Research Commission predicts a *slowing down* of the sea level rise. The increasingly erratic James Hansen is now talking about an eighty-foot rise in a few hundred years, which means we should already be seeing several inches a year instead of 1 to 2 mm. All sounds pretty "settled" to me.

Gulf Stream shutdown: Gore goes Hollywood and adopts the sci-fi *The Day After Tomorrow* scenario in which the ocean "conveyor belt" shuts off in the North Atlantic. While exciting, this doomsday scenario is unlikely. MIT professor of physical oceanography Carl Wunsch wrote,

"The only way to produce an ocean circulation without a Gulf Stream is either to turn off the wind system, or to stop the Earth's rotation, or both."[35] That is, so long as the Earth turns and the wind blows, we're okay. As Wunsch put it: "The occurrence of a climate state without the Gulf Stream any time soon—within tens of millions of years—has a probability of little more than zero." Seasoned Gore watchers know those odds to be close enough to a claim of "certainty" for his tastes.

Give me some feedback(s)*: Feedback* is a result that affects its own cause, and it can be characterized as *positive feedback* or *negative feedback*. If we think of a sports team, winning can bring in more fans, which generates more money, which allows the team to hire better players, which should result in more winning, bringing even more fans, and so on. This is a *positive feedback loop* because the effects of winning cause more winning—it *amplifies* the team's winning trend. *Negative feedback*, on the other hand, tends toward equilibrium. Think of a product: if manufacturers make more widgets, the added supply will drive down the price, thus discouraging further increases in production.

In climate, we see both negative and positive feedbacks. One positive feedback is that warming oceans let loose their CO_2, thus amplifying the greenhouse effect, contributing to more warming.

The global cooling alarmists in the 1970s loved to cite one positive feedback loop: as things got cooler, more snow and ice would cover the planet, reflecting more of the sun's rays (thus absorbing less), causing the surface and the water to cool even more. Thus, cooling leads to more cooling. Today they tell us that warming leads to more warming. Except

> "Available records dating back to 1897 and direct observation by the authors over a 4-year period [i.e., 1935–1938] indicate that Grinnell Glacier has been reduced to about half the size it was in 1900, and that the recession during recent years has been most rapid."
>
> **Gibson** and **Dyson** (1939) Bulletin of the Geological Society of America, vol. 50, 681–696.

for Gregg Easterbrook and *The Day After Tomorrow*, who argue that warming leads to a new ice age. The alarmist consensus is that a little bit of warming will cause *something* disastrous.

Al Gore says that melting ice caps reduce the planet's ability to reflect the sun's rays, allowing the world to absorb more warmth, which will mean more melting, and so on *ad infinitum*. Alarmists seem to think we used to have a Goldilocks climate—anything else would be too cold or too hot. Claim something is "unspoiled" by Man, and it's juuuuuust right.

Gore omits all other feedback loops. Clouds, for example, can be either positive or negative feedbacks, and there is still much debate over the nature and extent of the feedback role they play. Scientific literature includes many papers on negative feedbacks.[36] Ambiguity and uncertainty—what's been known as "science" for centuries, with minor interludes by the alarmists' "flat-Earther" brethren insisting on "consensus"—are however not helpful for Gore.

Extinctions: Gore claims that the current rate of extinction is one thousand times the background rate (that is the "natural" rate; yes, species do go extinct without poachers or polluters). In fact, the figures that Gore cites are admittedly simply based on a guess. In truth, we have documentary evidence of only a handful of extinctions in the last century.[37]

Gore's position comes from Norman Myers,[38] who based his position on Edward O. Wilson's studies, which extrapolate from extinctions on small islands. These studies, divining a number of extinctions from an observed loss of habitat, have since been shown not to apply to continental habitats. Myers argued that two million species will become extinct in the next fifty years. That means that about 40,000 species will go extinct each year.

There is no evidence for this assumption being true, but (like Paul Ehrlich's prognostications of doom) Myers's prediction is morbid enough to have earned him fame, riches, and high academic appointments.

If more than a few species a decade were going extinct, it seems that someone should be able to point to them. Contrary to this tiresome twaddle, it is well established that higher CO_2 levels will lead to much greater biodiversity. That's because plants use CO_2 to photosynthesize, and most classes of plants arose when CO_2 levels were much higher than today. Hundreds of studies published by the United States Department of Agriculture have established beyond a doubt that higher CO_2 levels will result in faster growth and more hardiness in nearly every class of plants. Several studies have already shown that the Earth is greening as a result. Generally this means plants and animals extend their outer ranges. Humorously, the greens now turn this into a negative, claiming that any plant we don't like—say poison ivy—will flourish in a CO_2-enriched atmosphere; lo, how the mighty scare stories have fallen. As with the BBC's big-tent alarmism, juxtaposing prognostications of global warming–induced calamitous killer storms and, oh yes, trouble for your garden too, the *Washington Post* weighed in with an ominous prediction of a world crowded out by the venomous vine, "Pumped up on carbon dioxide, vines strengthen their grip."[39]

"Even Mr. Gore qualified his statement [that 'the debate in the scientific community is over'] on ABC only a few minutes after he made it, clarifying things in an important way. When Mr. Stephanopoulos confronted Mr. Gore with the fact that the best estimates of rising sea levels are far less dire than he suggests in his movie, Mr. Gore defended his claims by noting that scientists 'don't have any models that give them a high level of confidence' one way or the other and went on to claim—in his defense—that scientists 'don't know. They just don't know.'"

MIT professor **Richard S. Lindzen**, "There Is No 'Consensus' on Global Warming," *Wall Street Journal*, June 26, 2006

Species migration: To follow this one, you need to remember the green rule: anything *new* is bad and a sign of worse things to come.

So, when the pine beetle appeared in Canada, it was another catastrophic effect of global warming. The immediate cause of this "invasive exotic species": fewer days with frost (somehow menacing your British garden, no doubt). Unfortunately for Gore's claim, the pine beetle has popped up in force all sorts of places in the past, for example Oregon's national forests, before DDT application wiped out the scourge. *Hail, DDT!*

Equally unfortunate is that in debunking some of Gore's claims, Tom Harris dared do something the greens rarely chance, turning to someone who might know what he's talking about as opposed to what he feels simply *must* be the case. Specifically, Harris queried Rob Scagel, forest microclimate specialist with Pacific Phytometric Consultants in Canada, who revealed that "The MPB is a species native to this part of North America and is always present. The MPB epidemic started as comparatively small outbreaks and through forest management inaction got completely out of hand."[40]

A forest microclimatologist versus a lifelong politician. Hmm. Anyone want to eat their hat over who's more qualified? By the way, for such sins the greens have initiated a campaign against Harris.

Storms. Again according to Dr. Robert Balling, "You will certainly not be surprised to see Katrina, other hurricanes, tornadoes, flash floods, and many types of severe weather events linked by Gore to global warming. However, if one took the time to read the downloadable 'Summary for Policymakers' in the latest report from the United Nations Intergovernmental Panel on Climate Change (IPCC), one would learn that 'No systematic changes in the frequency of tornadoes, thunder days, or hail events are evident in the limited areas analysed' and that 'Changes globally in tropical and extra-tropical storm intensity and frequency are dominated by inter-decadal and multi-decadal variations, with no significant trends evident over the twentieth century.'"[41]

It is inescapable for the curious that hurricane activity has been long understood to be cyclical. The cycle is on a long-anticipated upswing, as the greens warned us would happen. But it is not as aggressive as they would have liked. Witness the calm 2006 season. Regardless, the claim that warming will slightly increase hurricane intensity and/or frequency is a plausible hypothesis, if not universally accepted. Unsurprisingly, Gore exaggerates the hypothesis's size, scope, and acceptance. On the bigger point of 2005 storms and Hurricane Katrina specifically, Gore is just plain wrong, and the IPCC report finds no increasing trend toward more severe storms.

It goes without note in Gore's movie that a warmer world will, according to another hypothesis supported by physics and logic, lead to less severe and smaller storms because the temperature disequilibriums between the poles and the equator will be lessened. The historical evidence confirms that a warmer world is calmer and with fewer extremes. Brian Fagan's book, *The Little Ice Age: How Climate Made History,*[42] provides ample evidence for this—and a wonderful summary near the end of the first chapter of the shock that people experienced when the Medieval Warm Period gave way to the Little Ice Age. History does tend to repeat itself, both in climate and human behavior.

> ## A Film You're Not Supposed to Watch
>
> *The Greening of Planet Earth: The Effects of Carbon Dioxide on the Biosphere,* Journal of Environmental Education, 2000.

Vector-borne diseases. Gore asserts temperature increases will prompt mosquito activity and a rise in diseases. He is essentially just making this up in that it is wholly unsubstantiated. Unfortunately, highly regarded institutions such as the World Health Organization (WHO) push this line, and so Gore is not without authority to which he can appeal (if that authority also happens to be one heavily vested financially in the outcome, and after all funding dictates opinions, *right*?).

Underachiever

Al Gore's grades from college, as reported in 2000 by the *Washington Post*:

Natural Sciences 6 (Man's Place in Nature):	D
Natural Sciences 118:	C+

The disease claim is absurd. The planet could warm and malaria could spread, but the two would have nothing to do with each other (unless useless Kyoto-style policies exacerbated poverty, in which case we would find a once-removed correlation). Neither malaria nor the also-threatened dengue fever and yellow fever are actually "tropical" diseases. Russia's northern reaches have been hit with malaria outbreaks. Insect-borne diseases are not diseases of climate but of *poverty*. Moreover, they were common in non-tropical locales during the nineteenth century, when the world was by all accounts colder than it is today. Malaria was endemic not merely during the Little Ice Age but in the twentieth century in northern Europe (including Scandinavia, London, Edinburgh, and Riga), and northern North America (including Canada and Alaska, Toronto, and New York). Indeed, malaria was present along much of the east coast of America in 1882. Washington still had malaria in the 1930s. Europe was certified malaria-free in the early 1970s.

One thing is certain: temperature is not the cause of malarial activity. Instead, most experts on this subject agree that other factors are much more important in predicting future spread of these diseases.[43]

Malaria has in fact historically been present at high latitudes. A malaria epidemic in the Soviet Union in the 1920s, for instance, incurred 30,000 reported cases in the arctic port of Archangel, which at about 64° N is further north than the tip of Greenland.

Consider malaria's twentieth-century history in such well-known tropical locations, as detailed by the *Institut Pasteur's* Dr. Paul Reiter, former chief entomologist of the Centers for Disease Control's dengue fever branch:[44]

There was even some malaria in the Grampian Highlands of Scotland, which I assure you *never* feel tropical! And as you can see, the [15°] isotherm includes the south of Norway, much of Sweden, Finland, and way up in northern Russia. To the south of that line, every country was affected, many until quite recently. Holland, for example, was only declared malaria-free by the World Health Organization in the early 1970s. In the last century, outbreaks in the countryside around Copenhagen, Denmark, killed thousands. I can show you texts on the distribution of malaria in Sweden. Finland had a major problem until after the Second World War. Germany and France, too. Perhaps the worst problem was in Eastern Europe....Let me impress upon you how far north this problem went. In the 1920s and 1930s, the Soviet Union suffered dreadful epidemics of malaria, much more than any other country in the Northern Hemisphere. In the period 1923 to 1925, there were around 16 million cases, with 600,000 deaths. Archangel, which is at the same latitude as Fairbanks, Alaska, had 30,000 cases, roughly 30 percent of which were fatal. Think of it, malaria in Archangel. You need an icebreaker on the sea for six months of the year!

Malaria, then, is not a warm-weather disease but a poor-places disease. The Kyoto agenda poses a greater risk to spread malaria than does global warming.

Renewable energy and other proposed solutions. Renewable energy has a niche role as long as it is subsidized by government. Because it cannot survive in any meaningful way commercially without subsidies and mandates, that's about it. This raises a larger point about Gore's explicit and implicit message. The response he proposes is not commensurate with the magnitude of the problem he hypothesizes.

"No sense of proportion"

"Al is a radical environmentalist who wants to change the very fabric of America."

"He criticizes America for being America—a place where people enjoy the benefits of an advanced standard of living."

"He has no sense of proportion: He equates the failure to recycle aluminum cans with the Holocaust."

"He believes that our civilization, itself, is evil (because it is, in his words, 'addicted to the consumption of the earth.')"

From a 1992 **Democratic National Committee** memo by Jonathan Sallet to the Clinton-Gore campaign

Gore uses apocalyptic rhetoric to make the case for global warming as a colossal threat, but challenges the viewer with meaningless sacrifices. This, one should note, is precisely the same way the policy debates about "global warming" have proceeded, with no one daring offering anything that the alarmists, when pressed, could claim to have a detectable impact.

Gore does offer the gauzy rhetoric of demanding an effort equal to the Apollo space program and the Second World War combined, and says that the American people are capable of making such an effort. He also contradicts his own alarm by imparting the feeling that the calamity he prophesies is avoidable if we simply get off our butts, that the solutions are on the shelf and within reach but not employed due to a combination of sloth, unreasonable avarice, and, of course, dark and powerful greedy forces. *Reach for the moon . . . or merely the shelf at Home Depot with the more expensive light bulbs?* This makes ever more insufferable his decades-long invocation of Churchill (addictively raising the Nazi specter, as he and the environmentalists do so regularly . . . even demanding "Nuremburg-style trials" for those who disagree with them).

Churchill claimed that all he had to offer was his blood, sweat, and tears. Gore stops short of daring to call for sacrifice but instead promises an easy path for the individual—*let massive government intervention save you*—further indicating that his dreams of holding high office are by no means over.

That is, unless the public learns these very inconvenient truths.

Part IV

✳✳✳✳✳✳

MAKING YOU POORER AND LESS FREE

Chapter Eleven

❋ ❋ ❋ ❋ ❋ ❋

THE COST OF
THE ALARMIST AGENDA
MORE GOVERNMENT, HIGHER PRICES

Carbon dioxide emissions are the direct result of energy use. Energy use drives economic activity, and economic activity drives energy use. Therefore, even experts who accept detectable anthropogenic warming as reality leave no doubt that, regardless of any foreseeable technological developments, suppressing CO_2 emissions will restrict growth, destroy jobs, and diminish human welfare.[1]

Indeed, the countries that have significantly reduced their carbon emissions are, with only one exception, from the former Soviet Union and Eastern Europe. They did it the old fashioned way: economic collapse. The one exception involves the "one-off" political decision by the UK to dash to gas, a decision predating and unrelated to Kyoto.

Continued economic, technological, and population growth mean that energy use will increase. Some experts tell us to expect world energy demand to triple by 2050. There is no way to satisfy this sort of demand while simultaneously reducing CO_2 emissions as much as the alarmists demand.

We have a very simple choice. As a world, we could continue our march to improve the quality of life for all our fellow men, necessarily by providing electricity, heating, cooling, clean water, and transportation to the poorest people in the world. Or, we could follow Al Gore, stunt Third World development, and give up our own modern conveniences (including not just

Guess what?

❋ Global warming policies would increase our costs and reduce our freedoms, simply leaving us poorer to still deal with the uncertain weather that has always been there.

❋ Carbon dioxide taxes and rationing schemes are regressive: they disproportionately affect poor people and seniors.

❋ Renewable fuels cannot meet today's energy demands, let alone the world's growing needs.

automobility and labor-saving devices but energy-intensive, modern medicine), in the impossible pursuit of "stabilizing the climate."

Writing in *Science* magazine in 2002,[2] more than a dozen experts (many of them adherents of Manmade global warming theory), detailed how stabilizing greenhouse gas *emissions* without seriously damaging the economy is impossible at this time or in the foreseeable future.

Additionally, the Europeans and the Gores do not simply want to *persuade* us all to voluntarily sacrifice our families' quality of life for this crusade. They want to use government—from the local to the global level—and lawyers to restrict our freedoms and raise our cost of living, with obvious and significant human consequences.

These costs—in freedom, quality of life, and wealth (and thus health)—are almost universally ignored by the prophets of global warming and by the media. Gore has said this is intentional—that discussion of the costs of his plan of action ought to wait. By refusing to address these costs themselves, the alarmists enable their claim that the *only* reason anyone would deny their claims of scientific certainty and calls to action is because these skeptics are in the pay of Big Business standing to profit from catastrophic global warming. Meanwhile, it goes unnoted that these same alarmists are in fact widely supported by those industries standing to profit from the demanded "solution."

In truth, and despite the short-term profits envisioned by the greens' enablers, we *all* stand to lose big from their policies, which are simultaneously impotent (at "stabilizing the climate") and destructive (of wealth and quality of life).

Energy demand

The aforementioned *Science* article's authors, led by Martin I. Hoffert, predicted a tripling in world energy demand by 2050, and no foreseeable energy sources to satisfy this demand without continuing to produce

GHGs. In other words, Clinton's line that we're just too cheap or lazy to achieve the agenda that he chose not to pursue is pure slander. Gore's ersatz Churchill impression, promising to lead us away from the gathering storm he promises but without the blood, sweat, and tears, is equally absurd.

Increasing evidence shows us that all the popular "climate-friendly" alternatives—solar

> ## A Book You're Not Supposed to Read
>
> *The Real Environmental Crisis: Why Poverty, Not Affluence, Is the Environment's Number One Enemy*, Jack Hollander; Berkeley: University of California Press, 2003.

and wind power, biomass, nuclear power (for political reasons), fusion, fuel cells, ethanol, carbon capture and storage, efficiency upgrades—remain niche responses that cannot replace current energy sources without inflicting serious lifestyle degradations and cost increases. Instead, the Kyoto agenda requires that *all* energy use must be dramatically curtailed, because as the 2002 *Science* magazine article noted, "CO_2 is a combustion product vital to how civilization is powered: It cannot be regulated away."

To some climate change types, this is a bolder call to action—if the feel-good proposals of British Petroleum, DuPont, and John Kerry won't work, well, what the heck, then we need some drastic curbs on energy use. To regular people, this calls for skepticism—if the costs are this high, maybe we ought to take a second look. *Second looks* are not popular in our political system, or with environmentalists, whose ritual use of "we must act now!" also ritually betrays that their arguments don't withstand scrutiny.

Carbon rationing and Capitol Hill

The dominant "global warming" argument among even otherwise sensible U.S. lawmakers, when challenged on the wisdom of any given proposal, is "well, we have to *do something*." This mentality has set members of Congress on a years-long scramble to find what "something" is passable, regardless of its actual impact on the climate.

Let us put aside discussion of costs for a moment and note that this approach, intellectually bankrupt on its face, betrays the frustration among those with any understanding of the issue—be they alarmists or "skeptic": by "*do something,*" it is now quite clear that politicians mean "*be perceived as 'doing something,'*" not that they must actually "do something" to address the issue in substance. *We must "do something." This law is "something." Ergo, we must pass this law.* That is, politicians need to pass some laws that will do *something* (*anything!*), though not one of them is interested in policies that would even remotely affect climate change under any set of assumptions. (For more on the impotence of such policies, see the next chapter on Kyoto.) Policies that would affect climate change are generally wildly expensive (rationing energy) or just too com-

Kyoto vs. Kiwis

Kyoto taxpayer liability more than doubles, New Zealand says

"New Zealand taxpayers must pay NZ\$656 million to fund Kyoto Protocol carbon credits, the government estimated Wednesday, up from NZ\$313 million last year, due to an increase in expected emissions, higher credit prices and exchange rates.

While New Zealand had been estimated to release 36 million metric tons more carbon dioxide than it had allotted, the figure grew to 64 million after the government withdrew a proposed carbon tax and accounted for higher rates of deforestation. The latest figure is an excess of 41 million metric tons.

The current calculations also take a higher carbon price into account, NZ \$9.65 per metric ton instead of NZ \$6.

In 2004, New Zealand was expected to stay within its allotted emissions."

Brian Fallow, *New Zealand Herald*, October 13, 2006

mon-sense ("no regrets" policies such as achieving massive fuel savings by revamping outdated rules prohibiting airplanes from simply flying from Point A to Point B).

So, by "something," they clearly mean "nothing."

An (almost) humorous example of the legislative debate's fecklessness occurred in the Senate on June 22, 2005. On that day, while debating an energy bill, John McCain introduced an amendment "to provide for a program to accelerate the reduction of greenhouse gas emissions in the United States," carrying a price tag of $87.5 billion over a decade. The Senate rejected this measure 60–38.

Immediately afterwards, Democrat Jeff Bingaman introduced an amendment declaring: "It is the sense of the Senate that Congress should enact a comprehensive and effective national program of mandatory, market-based limits and incentives on emissions of greenhouse gases that slow, stop, and reverse the growth of such emissions." This measure—a non-binding resolution basically saying Congress should have passed what it had (wisely) rejected moments before—passed 53–44.

Translated: No, we won't do this. But we sure feel strongly that we ought to. The *Congressional Record* does not reflect how many members voted for this political prance with a straight face.

Exhibit A supporting the notion that Congress is more interested in luxuriating in the issue and the posturing that it affords than enacting any such painful prescription is that, to date, *not one* of the many proposals floated would have even the slightest detectable influence on the very same calamity purportedly underlying the legislation. Imagine claiming that you are so concerned about your child's inability to read that you insist on stocking his library with coloring books. That is the seriousness of our policymakers who most anxiously bemoan the lack of U.S. participation in energy rationing.

When climate-crusader Senator Bingaman touted a proposal of his at the Montreal Kyoto negotiation in December 2005 (a bill weaker than even the McCain-Lieberman "Climate Stewardship Act," which is a fraction of the insufficient Kyoto Protocol), I asked him at a press conference whether this represented a rejection of the alarmists' claimed threat—if he believed the threat, after all, he would propose something that would address the threat, not something even weaker than the watered-down version of impotent Kyoto. His remarkably candid answer, that he wouldn't be around when the real costs must (per the greens) be in place, blissfully manifests the desire to be seen as "doing something," not an actual passion to *do something*.

Perhaps Senator Bingaman realizes that *actually* "doing something" would impose greater costs than Americans could stomach, prompting them to ask questions about what benefit is gained in return. It might spur the politically incorrect debate that never was (yet is supposedly over and settled). In fact, doing nothing in the name of "doing something" would still impose those costs, as we see with the work of Senators Lieberman and McCain.

Cost/benefit analysis: Dividing by zero

The policy debate over "global warming" proposals certainly does raise moral questions, as proponents of Kyoto and McCain-Lieberman often point out. Yet the moral questions do not seem to be those that they intend. Quite simply, the entire logic for legislating is salvation from looming catastrophic Manmade global warming.

How, then, is it moral to propose action that will make an undetectable dent in this problem?

More important than the moral troubles of our McCains and Bingamans is the cost these do-nothing *do-somethings* will impose on regular people.

McCain's plan, for example, would cost us far more than the $87.5 billion in taxes we would pay over a decade to implement it. Competitive Enterprise Institute scholar Marlo Lewis reasoned in 2005 that the original McCain-Lieberman proposal (S. 139) would reduce Americans' Gross Domestic Product by $1.35 *trillion* over twenty years.

Lewis notes:

> Bingaman's plan would do less economic damage than the McCain-Lieberman bills, but it would also avert less global warming. Yet the total cost of any of these proposals is still huge. Whether the proposal is Kyoto, Kyoto Lite (McCain and Lieberman's original bill, S. 139), Kyoto Extra Lite (their pared-back versions, S.A. 2028 and S. 1151), or Kyoto-by-Inches (Bingaman's Climate and Economy Insurance Act), carbon-energy rationing is still a costly exercise in futility, as the following table shows.

Policy	(Best-case scenario)		(Worst-case scenario)		
	Tons GHG Reduced 2050	Warming Avoided 2050	Tons GHG Reduced 2050	Warming Avoided 2050	Cum. GDP Loss to 2025
McCain-Lieberman I	31,399	0.04° C	16,928	0.023° C	$1,354 billion
McCain-Lieberman II	21,285	0.029° C	11,320	0.015° C	$776 billion
Bingaman	5,816	0.008° C	3,163	0.004° C	$331 billion

GHG reductions are in million metric tons carbon equivalent.

At best, Senator Bingaman's plan would avert a hypothetical and imperceptible 0.008° C of global warming by 2050, and might avert only 0.004° C. Either way, the plan would not benefit people or the planet one whit. However, it would cost $331 billion in cumulative GDP losses. Don't the American people have better things to do with $331 billion?[3]

At least Senators McCain and Lieberman have the humility to declare theirs to be "a modest first step." One trembles for the economy when they imagine Mr. McCain's immodest second step.

Turning out the lights

Often, a global-warming prophet will try to win over the unconvinced by arguing, "even if you're not certain global warming poses this threat, shouldn't we take steps to prevent it, just in case?" This implies that they are proposing low-cost "solutions" to global warming. They are not.

Journalist David Freddoso outlined the steps we would need to take to comply with the Kyoto Protocol were Al Gore to become president and somehow get it ratified: "This is not a simple case of adjusting your thermostat by a few degrees, driving fewer miles this summer, or even buying a hybrid. Even gradual Kyoto compliance would require much more drastic action than that."

Freddoso constructed an admittedly silly scenario for how we could comply, but it shows the magnitude of the sacrifices Gore is asking us to make:

> According to the U.S. Department of Energy, the United States generated 5,802 million metric tons (MMT) of CO_2 in 2003. Naturally, this number has grown over the years as our economy has expanded. In 1990 we emitted just 4,969 MMT of carbon dioxide. If we had ratified the Kyoto treaty, we would have committed to cut emissions to levels 7 percent below that 1990 level—or to about 4,620 MMT.
>
> Can we cut emissions by that much? Sure we can. I'm looking at the Energy Information Administration's table of all fifty states' levels of carbon dioxide emissions. If we shut down all industry and electric generation in the fourteen "Blue" States (the ones that went for John Kerry in 2004) east of the

Mississippi River, then seize all automobiles, airplanes, and private land there, we would slightly overshoot the Kyoto goals.

As a more equitable solution, Freddoso proposes:

> In 2003, gasoline use in the U.S. accounted for 1,141 MMT or 20 percent of our total carbon dioxide emissions. If Congress acts today to outlaw the use of gasoline for all uses—automobiles, lawnmowers, generators, et cetera—we'd be within just 40 million metric tons of reaching our Kyoto goals. And that's great, unless you like being able to drive, or having food brought to your grocery store, or having ambulances and fire trucks that can respond to emergencies.

More likely than an outright *ban* of fuels and technologies that emit greenhouse gases, Congress would pass laws that *ration* energy use—which is exactly what a cap-and-trade scheme does. A more efficient approach would be increasing energy taxes, such as the gasoline tax. In Europe, despite per-gallon gas prices ranging from $5.00 to over $7.00,[4] their emissions from transport nonetheless continue to rise sharply.[5] The price difference between EU and U.S. gas prices is in taxes. In other words, despite Clinton administration promises that implementing Kyoto would require $1.90 gasoline,[6] we know from Europe's experiment—in a less expansive geography, less wealthy society, and one long designed to reduce dependence on personal automobility—that gasoline taxes would have to be increased exponentially in order to ensure gas at *over* $7.00 per gallon, in today's dollars.

Under a cap-and-trade scheme, in order to legally emit CO_2 for industrial (and someday soon, *personal*) uses, you would need to have your hands on CO_2 "credits." As with gasoline or other ration coupons in wartime, Uncle Sam would hand out these credits to power plants, factories, et cetera. When you emit CO_2, you "pay for it" with your credits, while paying for the fuel itself with actual money. If someone had more credits than he needed, he could sell his surplus to someone who needs more.

In the end, this means you have to *pay* for the CO_2 you emit, *on top of* paying for the energy (gas, electricity) itself, and the capital equipment required to turn it into a useful form (your car, your furnace, appliances, and so on). This raises the cost of making power or running a factory. That in turn raises the cost of the goods made at the factory, but more importantly it makes energy more expensive. When energy is more expensive, we all know, *everything else* gets more expensive. Also, when energy gets more expensive, alternative places to use it—say, Mexico—become more attractive. The popular term is "outsourcing," which, as we will see, is what Europe is already doing with its energy-intensive industry as a result of its "first step" Kyoto promise, which it nonetheless is spectacularly violating.

An alternative to buying CO_2 credits, or relocating, would be investing in low-CO_2 sources of energy, such as (certain) ethanol fuels, or intermittent solar or wind power (which aren't as low-CO_2 as advertised because, given their intermittent nature, fossil-fuel generation needs to be kept running idly as backup at inefficient, below-peak levels). Low-CO_2 fuels, except for nuclear power, are less efficient and more costly. If your power company uses windmills instead of coal, your costs will go up. If you run your car on ethanol instead of gasoline, your costs go up. (However, some of the added costs will be shifted to taxpayers who subsidize these technologies to the hilt—adding even more costs to regular people).

This cap-and-trade scheme wins praise as a "market-based solution." That is, the government creates a market for something which generally doesn't exist, by artificially creating scarcity through a cap on something or a ban on another thing. In practice this really means: "Someone gets rich off it" (probably DuPont, Archer Daniels Midland, and some Wall Street traders) and "a cleverly hidden tax."

For consumers, cap-and-trade feels just like a tax—it imposes higher costs. It is just a more expensive tax. In fact, a straight-up tax on energy would be far less costly (and therefore less harmful to the economy) than cap-and-trade schemes. The Congressional Budget Office is on record acknowledging "similarities between carbon taxes and cap-and-trade programs for carbon emissions. For example, both policies would raise the cost of carbon emissions, lead to higher prices for fossil fuels, and impose costs on energy users and suppliers of carbon-intensive energy."[7] The environmentalist group Resources for the Future counted that cap-and-trade is actually about four times as expensive to the economy as an energy tax designed to achieve the same outcome.[8]

Of course, passing a tax is also far more politically difficult. In 1993, Bill Clinton tried this more honest approach (an odd juxtaposition of words, I admit) proposing a "BTU Tax"—or a tax on fuel assessed according to the heat content, measured in British Thermal Units (BTU). Clinton was embarrassed, with his own party torpedoing the idea.

Kyotophiles have learned that the cap-and-trade quota system is easier for people to swallow. But a cap-and-trade scheme would cost the economy more than a CO_2 tax to accomplish the same reduction in greenhouse gas emissions. One cause of this inefficiency is that we would all be paying companies to do things that have no value—such as building windmills and burning coal and gasoline to turn corn into fuel. (We're already doing this, thanks to subsidies and mandates, but under CO_2 constraints, we would be paying more of them to do it more.)

It really ain't easy being green

Whether our government tries to reduce our greenhouse gases through trading schemes or taxes, the result is higher energy costs for all Americans. In 1998, Bill Clinton's Department of Energy examined the effects on the

U.S. economy if we were to ratify and comply with the Kyoto Protocol. It wasn't a pretty picture.[9]

First, the DOE made clear their assumption—necessary to keep the costs as "low" as they did—that we would be using *less* electricity in 2010 than we were in 1998. They predicted an electricity use reduction of 4 percent to 17 percent. This is not because we all would become more conscientious about shutting off lights, but because we couldn't afford to use electricity as much. This naturally ignores our increased energy use from technological developments such as Al Gore's internet—when someone shakes a mouse somewhere, someone else must burn more coal.[10] The report predicts electricity rates increasing by at least 20 percent and maybe even 80 percent over any non-Kyoto-related increases.

Electricity would be so much more expensive because Kyoto would force us to give up coal (which is domestically abundant and relatively cheap) and use more expensive fuels, increasingly imported as well, such as natural gas, to generate our electricity. In addition to switching to gas, we would have to "dramatically increase the use of renewables . . . particularly biomass and wind energy systems, which become more economical with higher carbon prices."

Don't get too excited. These fuels only "become more economical" in relative terms. In truth, Kyoto rules would make *real* fuels much more expensive, while feel-good renewables remain just as inefficient and only slightly more expensive. (Similarly, if the government added $200 tolls to the San Francisco Bay crossings, catching a Delta flight from Oakland International Airport to San Francisco International would "become more economical.")

In Kyoto's America we would cut our electricity use even more than the predicted 17 percent except that alternatives for heating our homes would be even worse. Natural gas prices would go up at least 25 percent or they could double—and that's the beginning. And how about that pioneering UK and their similarly dropping coal in the "dash-to-gas"? Unlike our coal, their gas is now running out, stranding them with energy price spikes and threatened energy shortages.

Gasoline prices would increase between 11 percent and 53 percent, on top of any price hikes future wars or Gulf hurricanes might cause. This, of course, would drive up costs of all goods shipped by truck—which is basically *everything* (the study estimates a 19 percent acceleration of inflation). You would not stop driving, however, as even Europe at $7.00 per gallon has proved. Instead, certain trips would be avoided, but, as intended, many folks would sacrifice their bigger (read: *safer*) cars for little ones, and significant choices would be imposed involving transport, work, and living. This is all to show that a meaningless "first step" involves sacrifice of convenience, safety, and quality of life. This is not, as Al Gore wants you to believe, a simple matter of pedaling your Schwinn to work once a week and switching out some lightbulbs.

The effect of Kyoto on the price of consumer goods would be mitigated by the fact that China, Mexico, and India—in fact, except for Canada and tiny New Zealand, the entire non-European world—are *not* bound by Kyoto's reductions, and thus would not experience higher manufacturing costs. That is, U.S. manufacturing would be offshored even quicker, and we would become even better customers of the Chinese, except that we

> **"T**here are no known technological options that exist today. Energy sources that can produce 100 to 300 percent of present world power without greenhouse emissions do not exist; either operationally or as pilot plants. New technologies will require drastic technological breakthroughs. Carbon dioxide is a combustion product vital to how civilization is powered; it cannot be regulated away. But carbon dioxide stabilization would prevent developing nations from basing their energy supply on fossil fuels."
>
> **Hoffert**, et al., *Science* magazine, November 2000

would have lower incomes to spend on their goods. If we have fewer steel mills here and more steel mills there, this provides no reduction in CO_2.

Kyoto, this Energy Department study says, would reduce our Gross Domestic Product by *at least* $77 billion (in 1998 terms, or $89 billion in 2005 terms) and by as much as $283 billion ($328 billion in 2005 dollars). That means our economy would be poorer by about $1,100 per person per year. The private accounting firm Charles River Associates estimated a cost of $225 billion per year.

A similar 2000 study by the DOE concluded that the average family's energy prices could be 18 percent higher had we ratified Kyoto. Businesses could be paying 30 percent more, and industry would be 19 percent higher. Those higher costs to business and industry, of course, would largely be passed on to the consumer—you.

It's been tried

Europe has tried "cap-and-trade" schemes on greenhouse gases, allocating quotas to certain industries. The result is massive energy cost increases, and industry (jobs) fleeing, with even the alarmists seeking to find a way to stop the capital flight.

Europe is harming itself so much that it has decided it needs to harm the U.S., too. Thus, the EU is considering a greenhouse gas *trade war*, looking at "taxing goods imported from countries that do not impose a CO_2 cap on their industry as a way to compensate for the costs of climate-change measures."[11] In short, they are contending that our government is subsidizing our industry by not taking control of the energy supply and parceling it out on political grounds.

Europe's suffering shows us what we would endure if we ratified Kyoto. Spain, for example, has closed at least three plants for failing to possess Kyoto permits.[12] At least one manufacturer has made good a threat to leave the country to escape Spain's blackouts and rationing of energy, Acerinox, which has sent recent growth to South Africa, and Kentucky (175 jobs, at its North American Stainless Steel subsidiary).

In the UK, where like elsewhere throughout Europe firms have to buy permits from firms in other member states, companies affected by the laws had to pay £470,000,000 (approximately $875,000,000) to comply in 2005, a year in which Britons' electricity prices coincidentally shot up by 34 percent.[13] In Germany, *Fortune* 500 energy and aluminum supplier Norsk Hydro Aluminum closed several production sites due to these higher costs arising from the emissions scheme which increased electricity prices.

If Europe actually pursues this agenda to a second stage, depending on the country, the damage will be considerable. Millions of jobs will be lost.[14] Recall that these costs do not, under any scenario, buy stable climate. Further, Europe's promised—but still unfulfilled—sacrifices are more than offset by the greenhouse gas increases in the countries not covered by Kyoto.

It may be foolish for the U.S. to jump on this sinking ship but, heck, it would make some people feel really good.

Europe admits that we are economically well-served by not signing on to their treaty, and they're angry about it. Now, the equally mad (take that as you wish) Al Gore and John McCain want to put us on the same page as Europe, thus protecting us from the jobs and wealth fleeing the continent.

Respected economic analysts Global Insight estimated in 2002 that meeting their Kyoto target would reduce Germany's GDP by 5.2 percent, Spain's by 5.0 percent, the UK's by 4.5 percent, and the Netherlands' by 3.8 percent.[15] Spain and Britain would each lose a million jobs, while Germany would lose nearly 2 million jobs, thanks in part to 40 percent increases in electricity and heating costs.

Greens disguise these costs by relying on unreliable econometric models to forecast little to no pain, and conclude that their policies would impose very little cost. The trick is to intentionally use a model that will only examine the impacts on a particular sector—say, energy production—and ignore the otherwise obvious consequences that the higher cost of energy and buying quotas have on users of energy. These models, called "sectoral" models, are the darling of the European Environment Agency.[16] This would be economic malpractice were it not officially sanctioned.

Similarly, here in Washington, it remains fashionable to ignore the costs of these climate-control schemes. In September 2006, the Congressional Budget Office purported to advise Congress on the costs of pricing carbon.[17] Although the very policies they claimed to examine were being tried in the laboratory of social experiment that is the European Union and its member states, the CBO mentioned the European experience exactly *zero* times, reminding us of the old economist joke, "Sure, it may work in practice . . . but will it work in theory?"

Sure, Kyoto-style schemes may be miserable failure in the real world, such as they are in Europe, but wouldn't they work on paper?

Lucky for Europeans, all of the *Sturm und Drang* notwithstanding, they are not reducing emissions as they promised under Kyoto.

Kyoto targets are unrealistic, attainable solely through a combination of cheating, as demonstrated in these pages and elsewhere, exporting growth and paying others under various Kyoto schemes designed to gain Third World buy-in (if not actual participation). Regardless of announced targets, thirteen (and possibly all fifteen) of the countries considered as "Europe" by Kyoto are on course to fail their particular emission as promised under Kyoto—which promises serve as the ostensible basis for their anti-American name-calling on "the environment"—including several that are spectacularly violating their promises.[18]

Blind optimists at home

Here in the U.S., advocates such as the journalist Gregg Easterbrook construct oversimplified and inapt syllogisms in pursuit of gaining public and policymaker approval for this agenda. For example, Easterbrook states that "global warming" is really an air pollution problem, that the U.S. history of regulating air pollutants is that the problem is regulated much more cheaply and more rapidly than predicted, and that once lawmakers "get their heads around" these truths they can therefore proceed to impose Kyoto-style policies without fear about the economy.[19] Seemingly aware of the argument's underwhelming persuasiveness, Easterbrook cheekily intimates that an economic boom might follow. As proof, he points out that the Clean Air Act's "acid rain" regulations were enacted in the early 1990s, and the decade was characterized by strong economic growth. *QED.*

Gotta love that logic! But one might use Easterbrook's same logic to point out that the one-degree warming of the past hundred years has coincided with an unprecedented increase in prosperity, standard of living and lifespan (as also occurred in the medieval warming). Quite a dilemma. Tough to get one's head around. Better to err on the side of heavy-handed government intervention in the economy and rationing.

One of the clever aspects of such tidy rhetorical approaches as that employed by Mr. Easterbrook is that they require much more time and space to debunk than to make. Regardless, given that this

"LONDON—When European Union officials created a market for trading pollution credits, they boasted that it was a 'cost-conscious way' to save the planet from global warming. Five years later, the 25-country EU is failing to meet the Kyoto Protocol's carbon dioxide emission standards. Rather than help protect the environment, the trading system has led to increases in electricity prices of more than 50 percent and record profits for RWE and other utilities. 'I don't suppose the environment has noticed the European emissions trading scheme,' said William Blyth, director of Oxford Energy Associates in Oxford, England, and a former International Energy Agency official who advises businesses on energy and climate change policy. The utilities and emissions traders, in contrast, 'have done very well.'"

International Herald Tribune, October 18, 2006

claim is becoming the mantra of warming alarmists (Easterbrook calls himself instead an optimist), remember that the "acid rain" program met with success (which he overstates) for reasons wholly inapplicable to Kyoto-style policies.

First, unlike Kyoto this program had an environmental impact and for far fewer billions than Kyoto-style policies involve. Any noticeable slowing of global warming would require a reduction in greenhouse gases beyond imagination given their central nature in our economy. Further, what is demanded is "thirty times" what the greens have had the temerity to put on the table.

Specifically, the "acid rain" issue dealt with two compounds, sulfur dioxide and nitrogen oxides, not six, as would be required of any GHG regime. These are pollutants, impure byproducts of the combustion process, while the principal focus of GHG regulations would be carbon dioxide (NO_x is also a GHG, though not the principal focus of Kyoto-style regulation), something that not only is essential to modern life but an *intentional* product of combustion: the more efficiently one combusts hydrocarbons the more CO_2 one produces.

Also, only a few activities gave off acid rain–causing sulfur dioxide, and so targeting them was easy. Greenhouse gases, on the other hand, come from everywhere—cars, factories, power plants, ethanol distilleries, your mouth, your nostrils, your cut flowers, i.e., the entire economy. Further, remember that some people in the world *are* trying the solution that Easterbrook proposes (Europe) and it is a miserable failure, with markets spiking and collapsing, energy prices skyrocketing, political considerations dominating allocation decisions, and cheating emerging on the critical data-integrity front, all resulting in massive costs to the economy with absolutely zero environmental benefit. It is no wonder the European Emissions Trading Scheme does not find its way into discussions such as Easterbrook's on *The News Hour*.

Finally, unanticipated deregulation of the railroads ensured that carrying low-sulfur coal from western U.S. mines to eastern utilities made all cost projections about the program obsolete. Had it not been for the option of shipping low-cost, low-sulfur coal from the Powder River Basin to the rest of the world, this program could easily have been a disaster. What major, almost flukish, policy achievement do advocates such as Mr. Easterbrook posit will serve that function in a carbon-constrained world? A revolution in human transportation? The Segway should do it, along with Gore's other prescriptions of eliminating the internal combustion engine, wrenching societal transformations . . . oh, and changing your light bulbs.

Stick it to the poor

We know the impact that energy costs have historically had on the U.S. economy.[20] Specifically recall the human toll taken when California last had to deal with rationed energy, also due to absurdly short-sighted and pandering politicians, enabled by greedy rent-seekers.

Who gets hit first and worst by such price increases: generally anyone on low and fixed incomes, such as seniors and the poor. As CBO stated in describing these options, "The price increases created by this policy would be regressive because lower-income households spend relatively more on energy."[21] If a middle-class family gets higher heating bills, they can cope by switching from turkey sandwiches to tuna. The poor family is already eating lentils and doesn't have much wiggle room when the Kyoto bill hits.

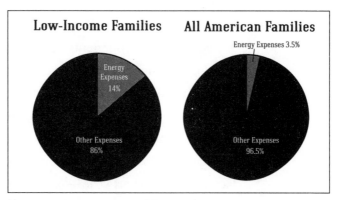

(Source: Department of Energy)

Elected Kyotophiles must explain to the public how, precisely, it is responsible to push a less effective, more greatly disguised scheme that is four times more expensive. In fact, how is it preferable to other options for any purpose other than hiding what it is you are really doing? To borrow the Left's language, do you really want to make Grandma choose between heating and eating?

Now, they will at some point doubtless figure out to say, *we can offer more low-income heating subsidies for those people*. Sorry, that won't float. You see, the basis for this scheme is that the rampant energy use of wealthy societies is immoral and, frankly, dangerous not just for this generation but the next. Why would the alarmists stand for a subsidy on such immoral activity? (Perhaps when old people or poor people give off CO_2, it isn't as bad? Regrettably, the environment recognizes no socio-economic distinctions between CO_2 molecules.)

No benefit

The question of whether the Kyoto Protocol would reduce warming is in fact an economic question as much as a scientific one. This is particularly true given its blunt instrument of rationing access to energy—let alone through the odious means of placing effective control over select sovereign nations' energy policies in the hands of the UN.

Even Kyoto's proponents acknowledge that despite accepting, for the sake of argument, Kyoto's underlying assumptions, and even were it implemented fully, universally, and perfectly, *it would have no detectable effect on temperature increase* (see next chapter). That is to say that even in theory and according to its proponents' best case (which has been already proven to be utterly unrealistic), Kyoto might avert—delay, in fact, by just six short years—projected future warming of an *undetectable few-hundredths-of-one-degree Celsius by 2050*.

This stunning impotence would however ensure that future growth in industry and agriculture would be outsourced from Kyoto's few covered countries to the vast majority of the world rejecting Kyoto's rationing. Kyoto exempts major developing nations and "top ten" emitters such as China, India, and South Korea. Mexico, Brazil, and others are free riders and happily intend to stay so.

This diagnosis of climatic meaninglessness would not change even were the United States to join up *and* comply with its terms. In fact, U.S. participation was one of the assumptions in the calculation.

It is critical to this most distorted of public debates as well as to "climate economics" to recall that it is this exempt majority of 155 countries—*not* just the United States and Australia—that refuse to accept any restrictions, now as well as in the future.[22]

Renewable boondoggles

The favorite line of the media and big businesses is that "new" technologies and "new" fuels can drive down our greenhouse gas emissions at relatively low cost. This is only true if decades of history and billions of dollars spent trying are wrong.

Recall the glum assessment from the 2002 article in *Science* magazine cited above:

> There are no known technological options that exist today. Energy sources that can produce 100 to 300 percent of present world power without greenhouse emissions do not exist; either operationally or as pilot plants. New technologies will require drastic technological breakthroughs. Carbon dioxide is a combustion product vital to how civilization is powered; it cannot be regulated away. But carbon dioxide stabilization would

prevent developing nations from basing their energy supply on fossil fuels.

(And those developing nations aren't happy about that prospect, hence their unified and persistent insistence on exemption from Kyoto.)

Alternative sources of energy such as renewables are still, after decades of promising cost-effectiveness soon, not yet cost-effective. They also come with environmental costs of their own.

Ethanol is a favorite new, environmentally friendly, domestic fuel here in America. Sadly, it is not new, environmentally friendly, or even necessarily domestic. It might not even be a fuel, actually.

Reduce Pollution, Increase CO_2

"Working in a tiny GM laboratory, [scientist Dick] Klimisch spent the next six years and $1 billion of GM's money to prove the skeptics wrong. Nicknamed 'Captain Catalyst,' he searched the periodic table until he found the right chemical combination to catalyze, or trigger, a reaction in the exhaust system of an automobile that rendered noxious emissions into **harmless gases**....

It was a few years earlier that the possibility of using catalytic converters as an anti-pollution device first attracted attention. In two years, Klimisch found by trial and error that a catalyst containing precious metals—platinum and palladium—retains enough oxygen when exposed to the high temperatures of engine exhaust to convert hydrocarbons into water vapor and **harmless carbon dioxide**, and carbon monoxide into carbon dioxide" (emphasis added; certain confusion of Lincoln Chafee disregarded).

Michael Weisskopf, "Auto-Pollution Debate Has Ring of the Past; Despite Success, Detroit Resists" *Washington Post*, March 26, 1990

Ethanol is alcohol derived from grain. Henry Ford knew moonshine could power cars, but we have always preferred real gasoline for many reasons. Gasoline is far cheaper than ethanol, even in recent times of oil price spikes. Ethanol only exists because of many government mandates and generous subsidies from sundry sources at all levels of its production.

Ethanol might not be good for the environment: The Clean Air Act would have outlawed it because it evaporates more easily than gasoline, causing more smog. The damage to soil from single-crop farming is probably more real than global warming.

Finally, the energy that goes into making ethanol—planting the corn, maintaining the fields, fertilizing the crops, irrigating them, harvesting the corn, shipping it, grinding it, distilling it, shipping the ethanol, and so on—uses more energy, from natural gas, coal, and gasoline, than the resulting ethanol yields, according to scientists Tad Patzek and David Pimentel.[23] In other words, making ethanol is possibly a literal waste of energy. (This point is very debatable, but the fact that it *requires* mandates and subsidies demonstrates that markets certainly don't think ethanol is a wonder fuel.) But it's hugely profitable for politically connected Archer Daniels Midland, which makes a quarter of the country's supply.

Corn squeezins are not the only eco-fuel that presents environmental problems. Veteran British environmentalist David Bellamy is leading the opposition to wind farms.[24] Hydroelectric dams, they tell us, are mean to fish. Currently the most cost-effective alternative to fossil fuel use is nuclear power,[25] which environmental activists continue to oppose in direct contradiction to their assertions that global warming is the gravest danger the planet faces.

The greater good

To the extent that the Kyoto agenda keeps poor countries in energy poverty, it will harm not only human welfare but environmental quality

as well. As Berkeley professor Jack Hollander shows, poverty, not affluence, is the environment's number one enemy, because poor societies lack
the wherewithal to protect the health and beauty of their surroundings.[26]

The key question for policymakers is what will do the most good for
human welfare. Climate alarmism is pitched for maximum fright potential: causing floods, storms, drought, insect-borne diseases, *etc.* As Danish statistician Bjorn Lomborg and Indur Goklany of the U.S. Department
of Interior, among others, have argued, even were Manmade climate
change as predicted (observational data already indicate that whatever it
is, it is nowhere the catastrophe the greens claim), we can save far more
lives by attacking these problems directly than by attempting energy suppression policies to control the weather.

Writing for the National Center for Policy Analysis, Goklany looks at the
curses climate change will purportedly bring upon the Earth—infectious
diseases, hunger, water insecurity, sea-level rise, and threats to biodiversity—and compares the cost of actually addressing those problems head
on versus attempting to mitigate climate change. There was no contest;
every scenario reveals that tackling extant threats directly now would be
considerably cheaper *and* more effective than imposing "climate change"
policies. "For example, meeting the emissions reduction targets of the
Kyoto Protocol will reduce the population at risk from malaria by just 0.2
percent in 2085. Investing as little as $1.5 billion in malaria prevention
and treatment would cut the death toll in half today."[27]

De-carbonizing our economy might make it harder to fight actual pollution. As AEI's Joel Schwartz documents, laws already on the books combined with the fact that people will, of their own free will, get new cars,
ensure that *air pollution* as a long-term problem is already solved.[28] (By
the way, Schwartz also notes in an email a May 2006 Gregg Easterbrook
op-ed in the *New York Times* that rings familiar given his climate change
claims. In this piece, Easterbrook claimed that certain regulations to
reduce auto emissions only cost about $100 per car; Schwartz details how

this is likely low by a factor of ten for cars built up to around the 2000 model-year, and several times more for subsequent models as EPA and California regulators tightened their standards. In short, the "regulate without worry for costs" optimism seems to be a bit of a default refrain not deeply grounded in experience.) Carbon suppression policies are a horribly inefficient way to curb air pollution—which by now the reader knows is not the same thing as greenhouse gases. For example, the U.S. Energy Information Administration estimated that Kyoto-level restrictions on CO_2 emissions from power plants would cost $77 billion but achieve only $6 billion worth of sulfur dioxide (SO_2) and nitrogen oxide (NO_x) emission reductions.[29] It would be far less costly to utilities, ratepayers, and the local economies to target the NO_x and SO_2—which are actually pollutants—directly and not use "global warming" as an excuse.

Poorer and no cooler

In short, climate is not inherently stable and neither are greenhouse gas concentrations, even if you take Man out of the equation (ask the next pontificating politician what the temperature would be without man, but don't expect an answer). Attempting to "stabilize" either one would cost trillions and would not stabilize climate. The jury remains deadlocked whether even massive interventions to stabilize *emissions* would have a climatic impact that is detectable with the most sensitive modern technology. This leaves open the question of how this much poorer world then deals with the ever-present, often severe and forever unpredictable weather, which will be with us in any event. The human impact of California's relatively mild energy rationing of just a few short years ago provides a bitter appetizer for the reality of an energy-constrained world.

My colleague Iain Murray reminded us of important facts in a November 13, 2006, opinion piece in the *Washington Examiner*, facts that Kyotophiles prefer you forget:

One hundred years ago, the average Westerner had an annual income equivalent to $4,000. A man could only work somewhere he could walk to; a woman spent much of her life performing back-breaking domestic labor. Medical science, while advancing, was still almost medieval in its practical application.

Much has changed in the last century, but in all cases the key to freeing us from these strictures has been widespread, affordable energy. A permanent flow of electricity has powered an explosion in wealth that has enabled millions to live long, fulfilling lives free from crushing hardship. The condition of life is no longer nasty, brutish, and short.[30]

We could try to make more of the world like Florida and less like Bangladesh, but you know where you'd rather ride out storms of equal severity. Physical and communications infrastructure—and economic resilience—requires wealth creation, which requires abundant, reliable energy. Unfortunately, at present the world remains energy poor. The Kyoto agenda seeks to make it more so. Run the numbers, take an honest assessment of feasibly available technologies, certainly inspect the motivations of the principal architect of this regime, and you will doubtless conclude that this is immoral. Passing ersatz measures in the name of climate salvation is equally so, for those who truly believe in the climate crisis.

In sum, you can't control the weather, but you can kill millions trying.

Chapter Twelve

✳✳✳✳✳✳✳

THE KYOTO PROTOCOL
A FRUITLESS, FAILED TREATY

The below chart says more than words ever could. It isn't my chart. It is from the European Environment Agency. The dark line is Europe's carbon dioxide emissions; the lighter line their total greenhouse gas emissions. The dotted line is what they promised the world—the basis, purportedly, for their actions in the UN on Iraq, for a pending trade war, for their name-calling, and for much European anti-Americanism.

Notice the year 1997, where the carbon dioxide emissions are at that point, and what's happened since. One might be tempted to say that global warming treaties aren't good for emissions.

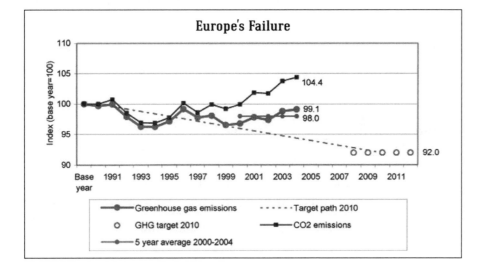

Guess what?

❊ Greens quietly insist that the Kyoto Protocol is merely one-thirtieth of what they want.

❊ Kyoto is already costing Europe dearly—which is nonetheless increasing its greenhouse gas emissions, faster than the U.S., no less.

❊ European officials have admitted Kyoto is about establishing "global governance" and is designed to "level the playing field" for businesses.

❊ The Clinton-Gore administration never sought to ratify Kyoto, for over three years after agreeing to it.

This is the Kyoto Truth, and apparently a painful one, given the denial in which we find so many policymakers. Europe promised their emissions would be down; they are up. Europe promised their emissions would be dropping; they are rising. Despite massive energy cost increases, increasing energy insecurity due to a manic drive to further depend on Russian gas, and remarkable, not to mention questionable, bureaucratic stunts, Europe is failing miserably at its promise to be the "world leader" in reducing emissions.

Yet they not only won't admit it, but the Europeans also ritually issue press releases trumpeting how Europe is "well on its way" and "on track" to reducing emissions as promised.

Up and rising, not down and dropping. What else could go wrong? Oh. Europe's emissions are rising faster than many others, for example, the United States'. Oooooh, that's gotta hurt. (Europe's emissions are rising three times as fast as the U.S.'s since 2000.)

And it does hurt. European governments' attempts to comply with Kyoto, while inadequate for reducing emissions, have had quite an impact on their subjects. Energy costs are climbing, hurting the poor particularly. As a final insult, even if they could curb their emissions, the effect on climate change would be minimal and probably indetectable.

Given this, it is high time to review precisely how we arrived at such a station, in which in the current, conventional wisdom is of a rogue U.S. "going it alone" in the face of our moral superiors' success.

Blame it on Rio

In 1992, an ambitious senator traveled down to Rio de Janeiro in Brazil for a fabulous party at expensive hotels full of like-minded people abusing expense accounts. This was not some decadent Gary Hart or Ted Kennedy romp—it was a campaign ploy by Bill Clinton's soon-to-be running mate Al Gore.

Gore went to the United Nations Conference on Environment and Development (UNCED) to complain that, while the rest of the world leaders were down in Rio talking about gas, President George H. W. Bush was somewhere else doing something productive. The elder Bush proved to have little resistance to such nagging, and gave in to the advice of White House chief of staff John Sununu, flying down to Rio and ultimately capitulating amid much abuse. Bush signed the original "global warming" treaty, the laboriously named United Nations Framework Convention on Climate Change (UNFCCC) or "Rio treaty," committing the U.S. to reduce its greenhouse gas emissions to 1990 levels by the year 2000. The Bush aides pushing the president into this deal stressed that it was a treaty with "voluntary" targets—in truth, it was a treaty that dignified the global warming sect of the man-as-agent-of-doom cult. This being an election year and with the Tennessee Scold hot on their heels, too, the U.S. Senate then rushed into ratifying the Rio treaty, unanimously, in embarrassing record time. In fact, among the about 190 parties to Rio, only the Seychelles and Mauritius acted faster.[1]

The fatal conceit of the Rio treaty was that it presumed to control nature. The treaty's statement of objectives read:

> The ultimate objective of this Convention and any related legal instruments that the Conference of the Parties may adopt is to achieve, in accordance with the relevant provisions of the Convention, *stabilization of greenhouse gas concentrations in the atmosphere at a level that would prevent dangerous anthropogenic interference with the climate system* (emphasis added).

Kyoto, Rio's successor, held the same goal. The problem with this goal is twofold.

First, media hype and deceptive Al Gore slide shows notwithstanding, greenhouse gas concentrations demonstrably do not determine temperatures. As discussed before, historically GHGs have gone up and down, but not in the same pattern as temperatures. To believe otherwise presumes that in the incredibly complex climate system all other things are and remain equal. Clearly, they are not and do not. Sometimes a GHG rise has preceded a temperature rise, and sometimes vice versa. Sometimes they move in opposite directions. It's a bit odd then for a treaty purporting to control "climate change" to claim to do so by moderating Man's contribution to greenhouse gas concentrations.

This raises the even more important point: *Man cannot dictate atmospheric GHG concentrations.* Throughout time they have been higher, they have been lower, all without our help. They are not static, and never will be. One might think the greens would show some more respect for Mother Nature, who, after all, produces 97 percent of greenhouse gases currently in our atmosphere by volume. Who gets penalized in the (common) event of massive natural releases causing a change in concentrations? In the event of a volcano eruption, increased release of gases from the oceans, or other natural activity, whose power plants must shut down or whose military must idle their engines?

Atmospheric concentrations of GHGs are no more inherently stable or subject to stabilization than climate, yet both of the "global warming" treaties claim stabilizing concentrations as their goal. The only thing we can (in theory) stabilize is *emissions*. As the chart above shows, this appears to be beyond the capabilities of even Europe.

That this "concentrations" goal has not been laughed off the pages of the establishment media make it clear that the media have never read Kyoto or given thought to its substance as opposed to its symbolism, theater, and politics.

These days, the parties pretend in their official statements that stabilizing *temperature* to two degrees above the temperature during the Little Ice Age is the agreed goal of Kyoto, equally absurd upon reflection.

The Senate's unanimous rejection

The U.S. failed to meet Rio's targets, just like many other nations.[2] "Global governance" types were soon agitating for another treaty—one that would *"legally bind"* nations to make strides toward that impossible goal of "stabilizing concentrations" of CO_2, nitrous oxides, and methane in the atmosphere, this time through emissions *cuts* from 1990 levels. As the agenda emerged and troubling noise emanated from the Clinton White House, Senators Robert Byrd and Chuck Hagel introduced Senate Resolution 98.

Byrd-Hagel offers a history lesson for Al Gore, John Kerry (who voted for it), Hillary Clinton, and all others claiming or implying that somehow President George W. Bush's opposition to Kyoto is aberrant. This bipartisan pair introduced the resolution in mid-1997, as the White House prepared for December 1997 talks in Kyoto, Japan. Republican and Democrat alike, senators sensed the Clinton administration was about to agree to a pact targeting the U.S. The resolution instructed the administration to not enter any agreement binding the U.S. to reduce greenhouse gases if the agreement did not include commitments by developing countries as well as developed and industrialized countries. Also, Clinton and Gore were to reject any treaty that would result in harm to the U.S. economy.

It passed 95 to 0 on July 25, 1997.

The resolution is unambiguous:

> Resolved, That it is the sense of the Senate that—
>
> (1) the United States should not be a signatory to any protocol to, or other agreement regarding, the United Nations Framework Convention on Climate Change of 1992, at negotiations in Kyoto in December 1997, or thereafter, which would—
>
> (A) mandate new commitments to limit or reduce greenhouse gas emissions for the Annex I Parties, unless the protocol or

other agreement also mandates new specific scheduled commitments to limit or reduce greenhouse gas emissions for Developing Country Parties within the same compliance period, or

(B) would result in serious harm to the economy of the
United States.[3]

This seemingly straightforward direction, the sort of "advice" called for
in the Senate's constitutional "advice and consent" role, was as we shall
see, apparently too opaque for the Clintonistas.

At about the same time, however, a little gas pipeline company out of
Houston was making its move up the ranks of the *Fortune* 500, buying
windmill and solar panel companies among other Kyoto-centric schemes,
expanding its enormous gas pipeline network that would skyrocket under
a carbon-rationing regime, and building (subsidized) coal-fired power-
plants in developing countries that presumably would not be covered by
a binding climate change treaty.

When the company's executive, one Ken Lay, paid a visit to the Oval
Office in August 1997, he leaned on Clinton and Gore to support the
Kyoto Protocol and offered very specific ideas as to what it must include.
Later he would pen op-eds in favor of GHG rationing. He was joined in
his meeting by British Petroleum chief Sir John Browne (now Lord
Browne), who was cobbling together a similar business plan he would
misleadingly dub "Beyond Petroleum."

By December, however, talks in Kyoto were going nowhere due to U.S.
negotiators taking the Senate's admonition seriously. Al Gore decided to
change that. With this corporate backing, and despite the unanimous signal from the Senate that agreeing to such a protocol would ensure it was
a dead letter, Gore boarded a plane to burn about half a million gallons
of kerosene[4] for a seventy-two-hour whirlwind from Washington, D.C.,
to Kyoto, just to instruct U.S. negotiators to demonstrate "increased
negotiating flexibility."[5] In English, that means *give in to what Europe*

wants. Consequently U.S. negotiators accepted an agreement that would bind the U.S. and Europe and very few others, leaving most of the world off the hook. Also, it clearly would "result in serious harm to the economy of the United States." Senate unanimity be damned.

What about those developing countries which per the Senate were to bear equal pain as the U.S.? Their sole obligation under Kyoto was to produce an annual report along the lines of "What I Would Do if I Had Made the Same Stupid Promise You Did." This was to be paid for by various funds established by Kyoto to be funded by you and me. As icing on the crepe, the ever-demanding (of others) French refused to promise to reduce emissions from the agreed baseline levels.

Bill Clinton waited a year and then signed the treaty, at the deadline for doing so and certainly with no Rose Garden ceremony. In fact, he so distanced himself from what is now a signature glory of his post-presidency, that he delegated it to an acting functionary at the UN, named Peter Burleigh. As any student of the Constitution knows, the president does not have the authority to bind the U.S. to a treaty. This requires approval of two-thirds of senators present. Considering not only Kyoto's premise being inherently flawed, its terms relatively punitive on the U.S. compared to Europe (and certainly the rest of the happily exempt world) and that that very same chamber, just months before, had preemptively said "over our dead 'world's greatest deliberative body,'" Clinton wisely spent the remaining three-plus years of his presidency *not* sending it to the Senate for ratification.

Keep this in mind next time you hear someone talking about how George W. Bush "rejected" Kyoto. Bush couldn't sign Kyoto; that can only be done once. He could seek ratification or, like Clinton, not. By any logic, therefore, if Bush *rejected* it, so did Clinton-Gore. More on that later, but now it's time to examine this treaty, whose praises are sung by the unwieldy choir of Gore, Lieberman, McCain, the entire media, a good chunk of the *Fortune* 500, and all of the European elite.

Kyoto's flaws

Kyoto as agreed in 1997 requires a handful of "developed" countries to limit—not necessarily reduce—certain gases to some level compared to 1990, specific to each country and by date certain (averaged over the five years of 2008–2012).

Clinton-Gore agreed to Kyoto terms allowing Europe to create a collective "bubble" under which most European countries caught a free ride on previous reductions made by Germany and the UK—reductions made *preceding and completely unrelated to Kyoto*: 1) the UK's "dash for gas" whereby Margaret Thatcher aimed to free the country from its strike-prone coal miners, and 2) Germany's shutting massive amounts of old, inefficient East German production capacity when the countries reunified. These are Europe's "Kyoto reductions."

Europe as a whole promised to reduce its GHG emissions to 8 percent below their 1990 levels, and the countries then got together to divvy up their reductions in a "burden-sharing agreement." In addition to the UK and Germany, three other countries actually promised to pull more than their share—at which they are nonetheless spectacularly flopping—while the rest of the EU would do its part by *increasing* emissions but within limits (up to 27 percent *above* 1990).[6] It is in this quiet arrangement that we find Europe's supposed world leadership in GHG emission reductions. Can you locate this reality in the laudatory (to the EU) and sneering (at the U.S.) media coverage of Kyoto?

The above explains why European think tank expert Christian Egenhofer says: "[Europe's] targets were the easiest....You could say that the US did a very bad job at negotiating [in Kyoto] while the EU did a good job."[7]

Kyoto also established a "cap-and-trade" system under which companies and countries may meet their quotas either by reducing emissions or—since that's obviously not happening—buying "credits" from others who regress economically or, against the odds, actually reduce emissions while continuing to grow (no such parties exist outside of the two unrelated exceptions cited). This idea was insisted upon by the U.S. in Kyoto. Europe opposed it bitterly, all the way through negotiations in late 2000. At that point, seeing its own numbers, Europe's tune changed. When Kyoto went into effect in February 2005 the streets of Brussels were festooned with decorations proclaiming the singular European achievement of an "Emissions Trading Scheme." As we shall see, this is Al Gore's (unwitting) revenge for the EU taking the U.S. to the cleaners.

Despite the advantages inherent to Europe in Kyoto's design, and despite the U.S. economy growing nearly three times as fast as Europe's, Europe's GHG emissions have increased *far faster than the U.S.'s*, and far faster than it had permitted itself under Kyoto.

Further bad news is that after nine years and ten negotiations, no one new wants to join. The vast majority of the world is exempt and intends to stay that way. Kyoto has been dead since its arrival.

The better news for Europe is that Kyoto, although touted as "binding and enforceable" is currently no such thing. At the December 2005 Kyoto talks in Montreal the Saudis proposed an amendment to make it binding. Europe refused. As such, the ineffective Kyoto is no more binding than the voluntary programs in the U.S. that Europe so bitterly declaims. Kyoto is useful for one purpose only, as a totem in the anti-Bush struggle.

Impotent Kyoto

For all the agonizing about Bush's "rejection" of it, you would think Kyoto was something that would actually help ward off global warming. It isn't.

Begin with the basic fact that the treaty professes to control something that no treaty can control: atmospheric greenhouse gas *concentrations*. Unless the United Nations can suppress volcanic, oceanic, or other natural releases (or absorption) of GHGs, there is nothing in the treaty to prevent wild fluctuations of greenhouse gases. The Protocol has no enforcement mechanisms to prevent trees, grasses, flowers, shrubs, and other plants from dying, or swamps from belching or cows from farting—and remember, natural processes account for nearly the entirety of the greenhouse effect.[8]

Stop Global Warming So We Can Keep Our Large Trucks Full of Oil Coming

Global warming isolates Canadians in far north

TORONTO, Ontario (Reuters)—Aboriginal communities in Ontario's far north are becoming increasingly isolated as rising temperatures melt their winter route to the outside world and impede their access to supplies.

❄❄❄

During the coldest months between January and March, "winter roads" are cleared on the frozen network of rivers and lakes to let trucks deliver bulk supplies like fuel and building materials.

❄❄❄

About 20,000 people live in the remote reservations and rely on winter shipments of heating oil, gasoline, and diesel fuel to power generating stations. The fragile ice has forced them to hire more trucks to carry lighter loads.

CNN.com, November 13, 2006

[Because clearly aborigines in the "far north" of Canada really don't want to be isolated.]

Even if we buy the European Union's sudden, unexplained switch to claiming that their Kyoto promise is now one of preventing 2° C *warming*, it's an impossible task. Remember, the planet has been cooling and warming for millennia, regardless of man's actions. All of the keening over modernity in the world will not change the fact that while nature may possess a global thermostat, Man does not.

Finally, let's imagine that Kyoto aims to control greenhouse gas *emissions*—not concentrations, and not a specific temperature, but solely anthropogenic emissions—a theoretically attainable goal. Even on this far more modest goal, Kyoto wouldn't do nearly enough to make a detectable dent in the global warming the alarmists are predicting (and that they are predicting will be catastrophic).

Not only does this treaty or any other do nothing to address "natural" greenhouse gas emissions (nearly all GHGs), it leaves alone most of the world's *Manmade* greenhouse gas emissions. Only thirty-five countries have agreed to cap their emissions—mostly big guns like Iceland, Belarus, and Slovakia—while more than 150 countries are exempt—including "top ten" emitters China, India, and South Korea, as well as Mexico and Brazil among others.

China and India are among the few parts of the "developing" world that are actually developing. That means they are industrializing, and thus increasing emissions. By 2009, China will be the planet's top emitter of greenhouse gases, according to the International Energy Agency's estimates. Kyoto imposes no limits on China's emissions and China insists it will stay exempt for at least our lifetime, during which period it will add a new coal-fired power plant every five days.

Poor countries are poor in no small measure because they are energy poor. Some two billion people on this planet have never flipped a light switch. China, for example, understands that a prerequisite of economic growth is getting enough energy for its people, which is why it is investing heavily in coal and nuclear power. These nations will never free their people from the drudgery and back-breaking labor we escaped last century if government denies them access to affordable, largely carbon-based fuels. This is why they reject Kyoto's rationing. They're already experiencing insufficient energy, and that is why they are poor. We have adequate energy, barely for now and with (also for now) only occasional exceptions, which is why our standard of living is so high, life spans are increasing, hunger is abated, and so on.

Given the many holes in Kyoto's purported assault on warming, it is little surprise that a scientist declared that the treaty would do almost nothing toward that end.

Thomas Wigley of the National Center for Atmospheric Research asked what effect Kyoto would have on warming.[9] He examined many different scenarios, and concluded: "in all cases, the long-term consequences are small." Wigley—firmly entrenched in the alarmist camp, mind you—estimates that in 2100, global-mean temperatures will have risen just over 2 degrees Celsius in the absence of the Kyoto Protocol. Wigley says that complying with Kyoto would reduce global warming by 0.08 degrees C—probably the difference between your right arm and your left arm right now. That is, with or without Kyoto, temperatures rise a little less than 2° C (just as looking out the window at the past three decades' warming would tell you we should experience at most 1.7° C, as explained in these pages).

Wigley also examines situations in which countries make bigger sacrifices than those purportedly required by Kyoto but still concludes: "The rate of slow-down in temperature rise is small, with no sign of any approach to climate stabilization. The Protocol, therefore, even when extended as here, can be considered as only a first and relatively small step towards stabilizing the climate. The influence of the Protocol would, furthermore, be undetectable for many decades." As ringing an endorsement of Kyoto as you'll ever see.

Would Kyoto ward off the flood of waters upon the Earth prophesied by Al Gore as punishment for the wickedness of Man? Wigley writes:

"Sea level rise reductions accrue even more slowly than warming reductions." Sea level would, according to Wigley, possibly be eight millimeters to twenty-one millimeters lower in 2100 than they would be absent Kyoto. He points out, "the prospects for stabilizing sea level over coming centuries are remote, so it is not surprising that the Protocol has such minor effects." Given that sea levels, like climate, are demonstrably *not* stable throughout history, but fluctuate with the ice-age cycles, this is a statement we should all be able to agree on. Consensus!

Again, if European governments continue hiking gas prices and heating prices, raising taxes to subsidize renewables, and losing jobs to China and India until 2100, they *might* just save eight to twenty-one millimeters of shoreline.

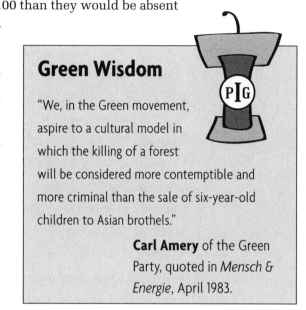

Green Wisdom

"We, in the Green movement, aspire to a cultural model in which the killing of a forest will be considered more contemptible and more criminal than the sale of six-year-old children to Asian brothels."

Carl Amery of the Green Party, quoted in *Mensch & Energie*, April 1983.

Europe's embarrassment

Amid all of Europe's scolding of the U.S. for not ratifying Kyoto, you might not notice that they are failing to live up to the treaty. Despite more than a decade of coercive policies and energy suppression in the name of "global warming," and their staggeringly high energy taxes, *Europe is not curbing its CO_2 emissions.*

As demonstrated in the chart opening this chapter, since Kyoto (1997), Europe's carbon dioxide emissions have *increased* dramatically. Since 2000 they are increasing three times as fast as America's. Remember, Kyoto mandated that Europe collectively reduce greenhouse gas emissions to 8 percent lower than the amount they were emitting in 1990.

When you look at the individual countries' performances, all but two of them have failed and project to continue to fail to meet this target. Before examining the data, it's important to remember the way the Europeans cleverly designed the treaty and its targets, or "did a good job negotiating in Kyoto" as their top policy expert says.

It was a very conscious, adamantly insisted-upon decision by Europe in *1997* to pledge instead to reduce their emissions relative to 1990—a baseline that just happened to afford them several advantages. Not least among these was that Europe was emitting fewer GHGs in 1997 than they were in 1990. Again, the reduction over the first part of the '90s was not due to any newfound sensitivity to climate change, but to two political decisions that had nothing to do with "global warming" (which had only been hatched in public in 1988, prior to which the mania was of course "global cooling").

When the United Kingdom, under Margaret Thatcher, eased government control of the energy sector, the new private energy companies made a famous "dash for gas" that enabled them to generate electricity far more cheaply, in part by scuttling overpriced unionized coal miners. Thatcher had been bedeviled by the serially striking coal miners in the UK holding their economy—and her economic reforms—hostage. Naturally, she had a strong role in seeing this transition come about. This steady shift from coal to nuclear power and the less carbon-intensive natural gas lowered Britain's greenhouse gas emissions throughout the 1990s.

In Germany, the Berlin Wall fell in 1989; East and West Germany reunified on October 3, 1990. In the words of an analyst for the center-left Brookings Institution, "Germany's reduced greenhouse emissions largely reflect a geopolitical quirk—it shut down wildly inefficient plants it inherited from East Germany."[10] It was that simple: shut down Soviet factories and shift the production to newer capacity, and declare the false *Wirtschaftswunder* ("economic miracle") of having grown economically while reducing GHGs!

The European Environment Agency puts it this way:

The main reasons for the favourable trend in Germany are increasing efficiency in power and heating plants and the economic restructuring of the five new *Länder* after the German reunification. The reduction of GHG emissions in the United Kingdom was primarily the result of liberalizing energy markets and the subsequent fuel switches from oil and coal to gas in electricity production and N_2O emission reduction measures in the adipic acid production.[11]

These two political actions, one driven by a demanding union, the other by communism's collapse, led these countries to reduce their CO_2 emissions. From 1990 until 2003, according to the UK's Office for National Statistics, the UK cut its total GHG emissions from the equivalent of 218.6 million tons of CO_2 to 192.6 million tons, an 11.9 percent reduction.[12] Most of that reduction was due to less carbon-intensive manufacturing and power generation. But the dash for gas was merely picking low-hanging fruit, and the UK's GHGs essentially flatlined around 1999, and indicate a slight uptick over the last few years.[13]

In late 2002, the Paris-based International Energy Agency issued a press release thanking Germany for helping save Europe from embarrassment:

> In 2000, Germany was only about 2 percentage points from its target under the EU "burden-sharing" agreement and provided the bulk of European Union reductions in emissions between 1990 and 2000. These reductions made it possible for the EU as a whole to meet the stabilisation goals of the UNFCCC, set in Rio in 1992.[14]

So if the other European nations can just find a communist-controlled half of their country with backward industry, shut it down, and send that region into an economic tailspin—*voila*! Europe will have met its targets. Almost humorously, Denmark's Socialist environment minister

Sven Auken presumed his native land could do just that, insisting on matching whatever promise Germany made in the "Burden Sharing" negotiations yet without the luxury of shutting down (hardworking) Jutland. As a result, Denmark projects that it will be "Europe's biggest Kyoto violator." This is only due to an absurd political promise, of course. Denmark's emissions are projected to increase 16 percent or more over 1990 by 2010, which is child's play compared to Spain's projection of being 48 percent above 1990 by the benchmark year of 2010, Portugal by an astounding 72 percent, Greece by 42 percent, and Ireland by 29 percent, among other violators increasing emissions faster than Denmark, symbolic as the home of Europe's Environment Agency.[15]

Europe is struggling with this reality. Though Spain formally reported this number to the United Nations as required, it is fair to say they issued no press releases about it. When I emphasized the failure, publishing Spain's own chart, during a March 2006 speech in Madrid, the Spanish environment ministry refused to confirm that any such figures—publicly available on the UNFCCC website—existed.

The European Union knows no such Iberian modesty as to refuse the opportunity to issue press releases, no matter how embarrassing the underlying truth. Consider the difference between its press releases and the actual emissions it reports. Both are released publicly with the full and rewarded confidence that no reporter will touch the actual report:

> **"Climate change: EU on track to reach Kyoto targets, latest projections show":**
>
> The EU is well on its way to achieve its Kyoto Protocol targets for reducing emissions of greenhouse gases *on the basis of the policies, measures and third-country projects already implemented or planned. . . .* The latest projections from member states indicate combined EU-15 emissions [reducing] to 9.3% below 1990 levels by 2010. This clearly fulfils the 8%

reduction target from 1990 levels that the protocol requires the EU-15 to achieve during 2008–2012.

EEA press release, December 2005, announcing
annual numbers (emphases added).[16]

The actual report, however, were a journalist to dare read past the press release, says: "*Even with* planned *additional* domestic policies and *measures, the target will not be reached.*"[17]

To learn this truth, one would have to roll up one's sleeves and read all the way down to the *first page* of the report's *first section*, in the *very first paragraph*. No reporter did.

Further, "The target will only be attained when Kyoto mechanisms are taken into account." While other tricks are actually required to reach the target, this report was really saying that European nations are *not* in fact reducing their menacing, hurricane-causing GHG emissions, but they do plan to pay Poland and the Czech Republic (whose post-communism economic restructuring provided plenty of low-hanging fruit), among others, in a wealth transfer scheme to buy GHG "credits."

> Bill Clinton's acting UN ambassador signed the Kyoto Protocol on November 12, 1998. Clinton relinquished the presidency at noon on January 20, 2001. That gave him **801** days in which to lift a finger in favor of Kyoto's ratification. No such finger was lifted.

So, the fifteen countries who were part of the EU when Kyoto was drafted (the EU-15, they are called), were by 2004 emitting more CO_2 than they have at any point in their history, with increases five of the seven years since they made their Kyoto promise.[18] When you consider *all* greenhouse gases, the EU-15's emissions are, as of the most recent figures (2004), down almost 1 percent from 1990, but rising. This is nonetheless 7 percent higher than the goal they are supposed to reach by 2010 under Kyoto, and going the wrong way. Meeting this goal

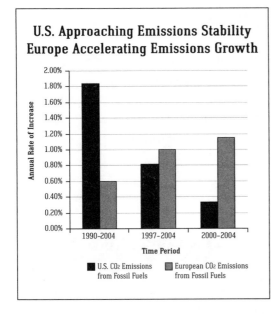

U.S. Approaching Emissions Stability
Europe Accelerating Emissions Growth

might be possible were they on the decline, which they are not. UK chancellor of the exchequer Gordon Brown humorously trumpeted that Europe had reached its target several years early. Given that their CO_2 emissions are rising, this is the last time they will see levels down where they promised they would be. By this logic, the U.S. achieved its 1990 emissions levels seventeen years ago, right around 1990.

Using any benchmark but one, the EU-15's CO_2 emissions have gone up and faster than the United States'. The chart at left details European CO_2 emissions from 1990 to 2004.[19]

Only when they use 1990 as the baseline can they claim superior performance to the U.S.—again for the simple, unrelated reason of the UK and German circumstances described above.

At some point, one might imagine a chastened Europe would dial the rhetoric down a bit—but not as long as the unquestioning media give them, and their enabling American politicians, a free ride on the issue.

Still, Europe regularly boasts about its spectacular Kyoto leadership, while the numbers tell the opposite story.[20] These are not secret numbers. They are public information, but journalists seem to have trouble finding them.

While Europe promised their emissions would be down, they are up; though they promised emissions would be dropping, they are rising. Meanwhile, the EU engages in name calling and blames the U.S. for causing hurricanes. Here's an even more inconvenient fact for Europeans, Al Gore, and John McCain: the U.S. performs much better at curbing emissions.

Like Europe, our emissions are up since 1997, and, because we did not have the benefit of the East German or UK anomalies, our greenhouse gases are also higher than they were in 1990. Since Kyoto was agreed in 1997, however, we are performing better than the EU-15 (as shown by the middle pair of bars in the above graph). Both we and they are increasing emissions, but in the past five years for which have data they are increasing *three times* more rapidly. Kyotophiles at this point in the discussion tend to change the subject to per capita emissions, hopping on the "wasteful Americans" horse they've beaten to death ages ago, and belied, below.

Bad science, bad math

Despite, or possibly even *because of*, Kyoto's inherent fecklessness and Europe's own failure to match its rhetoric with action, we are routinely bombarded with whines about a gluttonous United States population recklessly slurping energy to fuel its selfish ways while unfairly feeding the poison fruit of its lifestyle to the rest of the planet. The talking point perhaps most prominent among robotic greens worldwide is that the U.S. contains only 5 percent of the world's population but emits 25 percent of the planet's (Manmade) GHGs.

This is not only an exaggeration of U.S. emissions, but also an intentionally twisted way of viewing the U.S. contribution. It ignores the important fact that the 300 million Americans also contribute far more than their share of economic productivity to the planet at—this is what kills the whiny argument—a lower rate of GHGs per unit of economic production than the rest of the world. If you compare the U.S.'s contribution to the global GDP (made possible by our energy use), we are contributing *less than our share* of greenhouse gases. People are not just stomachs and feet that push gas pedals, but a set of hands and a brain. They are engaged in creativity and industry.

It is illogical to compare U.S. energy consumption with those huge swaths of the world that continue to live in deadly poverty, though that is precisely what the greens' "5 percent/25 percent" snivel does. This poverty persists in part because so many people are too busy scavenging their next meal or dose of medicine to do much else, let alone produce exports, remove bacteria from their water, keep themselves warm, keep their food cold, or partake in other reprehensible behaviors that pollute the air with CO_2. The world is not awash in energy, but suffering for lack of it.

Comparing countries' GHG emissions per capita is misleading because it assumes that poverty is the norm (of course we will have higher emissions than North Korea; we actually have electricity). This comparison also punishes productivity—an economy where each worker can produce only one widget a day will produce fewer emissions per capita than one where, thanks to automation, the worker is producing one hundred widgets per hour.

The saner measure is to compare emissions per contribution to global economic productivity. This measure, however, is not useful to the greens, because it doesn't make the U.S. look so bad. In fact, we emerge from this analysis as paragons of climate virtue.

We know that the rest of the world understands that rich countries are different from poor countries, and that one's influence and responsibility should be gauged not on the basis of population but contribution the global economy. We know this because the United States is dinged for a quarter of the United Nations' budget (22 percent of the operating budget and 27 percent of its "peacekeeping" budget). One complaint most certainly *not* heard on college campuses or among the Salvationists is that a country holding a mere 5 percent of the world's population carries a quarter of that particular institution's dead weight.

Conveniently, a quarter also happens to be the approximate U.S. contribution of the world's (Manmade) GHG emissions. According to the U.S. Energy Information Administration's figures, in 2003 the U.S. emitted

5,808 million metric tons (mmt) of GHG, out of a total global output of 25,162 mmt—that's 23 percent of the world's anthropogenic GHGs.[21]

These emissions are in return for what economic contribution? According to the CIA World Factbook,[22] 2005 Gross World Product (GWP) was $59.59 trillion. The U.S. provided $12.41 trillion, or 20.8 percent. Hmmm, 21 percent of the economy produces 23 percent of Man's GHGs. So the greens have a very small point, but when you consider that the American government is the most generous in the world, as are private Americans, that 2.2 percent becomes a wash.

The 20.8 percent figure results from measuring the GWP by the standard of Purchasing Power Parity (PPP)—probably the most useful and relevant standard, and certainly that accepted as the professional norm. Greens, in the context of global warming, instead insist on using the standard of Market Exchange Rates (MERs).

Using the green-demanded MERs, the GWP shrinks to $43.92 trillion, of which the U.S. is responsible for GDP $12.47 trillion. That's 28.4 percent of the productivity, for which it produces a disproportionately low 23 percent of the world's (manmade) GHGs! What this means is that, were the rest of the world to become as GHG efficient as the U.S., Kyoto's desired emission reductions would be obtained, several times over, and through

"'[T]he vast majority of Europeans would be delighted' if President Bush fails to win re-election on November 2. European animosity toward the incumbent began with his administration's rejection of the Kyoto Treaty on the environment,' Parmentier says, 'and grew over tensions caused by the war in Iraq.'"

Council on Foreign Relations, on-line interview with French foreign policy expert **Guillaume Parmentier**.

economic advancement, not restriction. That would certainly matter, were the underlying objective of Kyoto actually to simply reduce GHG emissions.

Bush and Kyoto

Europe's greenhouse gases are rising many times faster than ours. We emit less than our share of greenhouse gases. Still, Europeans seem to blame us for the climate catastrophe they say is already upon us. Specifically, they even blamed us for causing Hurricane Katrina. Seriously. Germany's then environment minister Jürgen Trittin stated in an op-ed, "The American president is closing his eyes to the economic and human costs his land and the world economy are suffering under natural catastrophes like Katrina and because of neglected environmental policies."[23] (Let lie the temptation to point out that Europe's performance might just point the finger at them for such storms.)

In the end, however, because they can't really point to performance, they point to intentions and words. The U.S. is ruining the planet because it has rejected Kyoto, they argue. GHG performance clearly has no place in this analysis. We can only exculpate ourselves by turning to the global greens and telling them "you're right." Clinton did that by signing Kyoto. Then, like Bush, Clinton proceeded to lift not one finger to seek Kyoto's ratification for over three years. But because of his signature, he is their hero. Unlike Bush, he told them, "you're right." That's what counted then, and still counts today. That's childish.

Equally juvenile is the specific claim (and Democrats join in on this one) that George W. Bush's unilateralism and opposition to science are threatening the planet and ruining the goodwill Europe used to hold toward the United States.

This, of course, is hogwash.

Remember, Bill Clinton delegated to his ambassador the task of signing the Kyoto Protocol, but never pushed nor even asked the Senate to ratify it.

He never even transmitted the treaty with a note leaving it to their own devices and counsel. It is true that the Senate had voted, in advance and unanimously, that the U.S. should not enter a global warming treaty that would damage the U.S. economy (Kyoto *would*) or that exempted most of the world (Kyoto *does*).

This required hard work by the administration, to change the terms of the treaty (instead, they signed it a year later), or change the minds of either the reluctant U.S. Senate, or the majority of the world that remained exempt. The day the U.S. agreed to the treaty, Al Gore said: "As we have said before, we will not submit the Protocol for ratification without the meaningful participation of key developing countries in efforts to address climate change." No such lobbying of the world was evident, and the administration never sought Senate approval.

So the Clinton-Gore administration's position was: we will sign the treaty, but not pursue its ratification.

It is difficult to distinguish this from the Bush administration's position. First, the executive's approval, in the form of a U.S. signature, is still attached to the treaty. Despite remarkably sloppy reportage to the contrary, President Bush never refused to sign Kyoto. He never had the option—it was already signed. Further, while Senate rules are ambiguous on the question of whether they require transmittal, the Constitution allows the Senate to vote up or down on signed treaties, with no mention of a formal "submission" or transmittal to them. Case law and the Constitution are clear: no court would impede the Senate from voting on Kyoto tomorrow. That means that at any point since 1998, including the time period when Tom Daschle was majority leader of the Senate and thus Democrats controlled the floor, the Senate could have held a vote to ratify the treaty, at most needing a rule tweak. They never have. The new Democratic majority could do the same. They won't. The Executive Branch's constitutional role in the treaty process is complete. However, given the End-of-Days rhetoric seeping from some of our senators, it seems remarkable that not one has sought to bring the treaty to a vote. It is of course so much easier to simply cast blame, particularly when the press allows one to toss about phony comparisons between administration actions and claims of withdrawals that never were.

Conversely, President Bush *could* move us further away from ratification by, in effect, unsigning the treaty. He did that with the Rome Treaty establishing an International Criminal Court.[24] He hasn't done it with the Kyoto Protocol. So, the Bush administration's position—not rhetoric—is indistinguishable from the Clinton-Gore position: our signature is on it, the Senate may work its will if they wish, but we will neither withdraw from it nor do anything to encourage its ratification.

Still, Bush gets pilloried for "rejecting" Kyoto. If Bush "rejected" Kyoto, so did Bill Clinton and Al Gore. The sole distinction is their respective rhetoric—Bush has been honest that he won't try to get it ratified—which is hereby established as what really matters.

Reuters wrote in 2005 that "Many international leaders have criticized Bush's refusal to sign Kyoto."[25] London's *Independent* similarly wrote that "President George Bush has refused to sign Kyoto."[26] The *St. Louis Journalism Review* parroted this line, writing of the "Bush administration's refusal to sign the Kyoto Protocol." On six occasions, the Associated Press has written about the Bush administration's "refusal to sign Kyoto," including one mention by the White House correspondent. The *New York Times* last November wrote that the U.S. "refused to sign" the treaty. Seeing how Bill Clinton already signed Kyoto, Bush can't sign it. (Should we expect stories about

Green Wisdom

"In short, if we can rise to the challenge, the permanent abolition of the wheel would have the marvelously synergistic effect of creating thousands of new jobs—as blacksmiths, farriers, grooms and so on—at the same time as it conserved energy and saved the planet from otherwise inevitable devastation."

Catherine Bennett,
The Guardian (UK),
2004

President Bush's "refusal to sign the Treaty of Paris" or his "refusal to sign the Declaration of Independence"?)

More careful—but just as dishonest—critics say that President Bush withdrew from a "Kyoto process." Having subjected myself to many of the meetings that constitute the "Kyoto process," I therefore need some explanation for the American hordes in pinstripes wearing State Department badges since 2001. Reporters even attend press conferences by the U.S. delegation obviously quite present and the same day quote in their stories UN-types demanding that the U.S. rejoin the process. Senate Foreign Relations Committee member John Kerry certainly should know as much, given that he approves the money spent to send our delegates. Under Bush the U.S. continued to send typically the largest delegation of any nation in the world to the Kyoto negotiations.

But unless we're there to tell the greens they're right, we aren't really there. We really should ask for our money back.

Prattling political polemics

The list of politicians who blame Bush's stance on the Kyoto Protocol for rampant anti-Americanism seems endless. As the Democratic Party's most recent standard-bearer for the White House, Senator John Kerry of Massachusetts is a fine case study.

After announcing his exploratory presidential bid, Kerry delivered two of his first major policy speeches premised largely on President Bush's mythical policy reversal on Kyoto. These were his speech on foreign relations at Georgetown's Walsh School of Foreign Service,[27] in which Kerry asserted that Bush's Kyoto withdrawal was reaping consequences for Middle East diplomacy, and his speech on the environment at Harvard's Kennedy School.[28]

Aides billed the Georgetown speech as a signature issue—the U.S. needs to play nicer with others. Kerry's timing was impeccably Gore-like for its dissonance from real-life world events, delivered as it was in the midst of a "record-breaking" and deadly cold wave. It was also wonderfully enabling, coinciding with the French and Germans escalating their coordination aimed at rendering the United Nations Security Council "impotent" (in the accurate phraseology of Secretary of State Colin Powell).

Kerry complained that the Bush administration's "high-handed treatment of our European allies, on everything from Iraq to the Kyoto climate change treaty, has strained relations nearly to the breaking point." (Later, his wife reiterated this ignorance, actually bemoaning Bush's *refusal to sign* Kyoto for the discomfort they were doubtless forced to confront at continental cocktail parties.)

Soon after his Georgetown speech Kerry again offered his Kyoto/Iraq relativism in an interview on *FOX News Sunday*. He suggested the Europeans would work with us constructively on battling terrorism if we gave in to them on battling affordable energy:

> This administration has broken the relationships that we had, the strength of those relationships. And there's a lack of credibility and a lack of trust. *This administration has never, ever fully offered these other countries the kind of partnership*, the kind of decisionmaking sharing, the kind of participation in the reconstruction, *the kind of participation in other issues that matter to those countries that actually bring them to the table.* I know how to bring these countries to the table. And there are some very powerful cards we have to play....The European countries have a huge stake in not having a failed state in Iraq, but they're not at the table today. The truth is, this president has failed in his conduct of diplomacy[29] (emphases added).

Kerry blamed France's misbehavior on Bush's Kyoto stance regularly. He even trotted this mythology out during the first candidates' debate after receiving the nomination.[30]

Bill Clinton stayed true to form in his July 26, 2004, keynote speech at Kerry's nominating convention in Boston, with a repetitive rant about how the U.S. had made its bed of angry allies by squandering "post–9/11" global goodwill by affirming the U.S. policy of not seeking ratification of Kyoto, *six months before September 11* (Bush made his position clear in March 2001).

This pair is not alone in their mindset that there simply are greater factors at play when the U.S. asserts policy decisions on sovereignty and national security than simply our own selfish notions of sovereignty and national security.

Possible future Democratic nominee Senator Hillary Clinton apes Kerry's argument that the U.S. "has not led but fled on global warming" and should now "rejoin our allies at the negotiating table."[31] Like most smart politicians, Hillary is reticent to invoke the K-word itself. Senator Clinton chooses to speak solely in terms of "climate change." Still, she makes clear her sympathy for the party line that under Bush the U.S. just walked away from something, reversing the course of events pursued under her husband's administration. For example, the following from a speech at the National Press Club in 2006 is typical of her rhetoric on the subject:

> [I]f we're going to *reassert our leadership* on climate change—which I think we should—we've got to deal with coal. And the first step is to take a mandatory cap-and-trade system, like that developed in the McCain-Lieberman legislation that I support, but obviously going out and trying to *reengage* the rest of the world in this issue.[32]

To review, the only U.S. president whose team ever "disengaged" from or walked away from the table as regards Kyoto is Bill Clinton. This disengagement occurred at the failed November 2000 talks in The Hague, when Europe sought to take advantage of a presumably desperate U.S. team of Gore acolytes as the Florida recount unfolded in the courts.[33] To avoid economic devastation to the U.S. while pursuing GHG reductions, negotiators had agreed to allow countries to use GHG *sinks*—forestry or other land use practices absorbing or not emitting atmospheric gases. The Europeans wanted to scrap this—only reducing emissions would do.[34] The U.S. up and walked away. The greens, at the time, quite rightly blamed Europe.[35]

The U.S. returned to the next talks in force, under George W. Bush, still facing and—this being their sin—*acknowledging* the unanimous Senate vote against such an agreement, as well as his openly expressed personal opposition. This also, by the way, belies the occasional defense of Bill Clinton vs. Bush on Kyoto, that Clinton at least sought to fix Kyoto's problems while Bush just walked away.[36]

Having hereby corrected the record, let us note that Hillary's mantra about "reengaging" and "reasserting our leadership" also betrays an ignorance or revision of history that so fulsomely emanates from Kerry, Daschle, Bill Clinton, and the rest. She also exposes sympathy for a president ignoring the Senate's clear pursuit of its "advice and consent" role. Hillary still views herself as a "co-president," as she bides time and builds a persona to run on her own, with consistent disdain for the minor constitutional impediment to executive authority that is Article I of that guiding document (the legislature), and other limits such as those found in Article II, Section 2.

No Blood for Oil! But Blood for Kyoto? *Oui!*

"One reason Washington's goodwill reserve had all but vanished [when seeking support at the UN for its Iraq stance] is that European countries pay a lot of attention to treaties....Nonetheless, early in its term, the Bush administration declared war on all outstanding international treaties. First he repudiated the Kyoto Protocol on the environment."

Former Clinton administration State Department official **James P. Rubin,** "Stumbling into War," *Foreign Affairs,* September/October 2003

Global governance

It is already apparent that the Kyoto Protocol isn't working. No party to Kyoto is reducing emissions since agreeing to the treaty, and the only ones who can claim emission reductions must a) invoke Kyoto's 1990 baseline for such a claim and b) rely upon reasons completely and inarguably unrelated to Kyoto. Not surprisingly, no one else wants to join the sinking ship. Even if it functioned as imagined it would make us poorer without making us cooler, and thirty Kyotos would without doubt be a cure worse than the alleged disease. Most climate change risks can be handled more effectively at less cost through wealth- and resiliency-enhancing measures.

The risks of climate change policy—poverty and coercive wealth redistribution—outweigh the risks that might be expected as climate continues to change. Bjorn Lomborg correctly asks why would anyone propose spending twice as much money as would address, and in great part solve, the world's existing problems, such as unsafe drinking water, real pollution, mosquito-borne diseases, and AIDS, in order to do nothing? The answer has everything to do with the real environmentalist agenda, and nothing to do with "climate change."

First, it's important to remember the words of Europe's then commissioner for the environment, Margaret Wallstrom (now EU vice president for, of all things, communication): "This is not a simple environmental issue where you can say it is an issue where the scientists are not unanimous. This is about international relations, this is about economy, about trying to create a level playing field for big businesses throughout the world." But the treaty doesn't target Mexico's, China's, or India's competitive advantage gained through lower environmental standards—those nations' emissions would not be curbed. Quite clearly, they would increase, due to Kyoto's inherent incentive to offshore energy-intensive activity from wealthy countries with stringent environmental standards to these laggards. Well, Wallstrom was making it clear—Kyoto was about hobbling *America's* economy to give Socialist Europe a fighting chance.

When considering European cries about Kyoto, it is critical to remember that the dream of the environmentalists is global governance. French president Jacques Chirac is blessed with the virtue of forthrightness. At the infamous Hague talks, Chirac called the Kyoto Protocol "the first component of an authentic global governance." There it is. Kyoto is the tool by which the Chiracs of the world can finally get some control over us unruly Americans.

Unluckily for Jacques, John Kerry's "global test" did not fly with American voters in 2004. Chirac also lost an ally in the U.S. Senate in 2004, thanks to the cruelties of the democratic process. Consider this exchange in 2002 between then Senate Majority Leader Tom Daschle and FOX News' Tony Snow:

> **SNOW:** You were harshly critical the other day at the Bush administration's foreign policy. Once again you said, "I don't know if we've ever seen a more precipitous drop in international stature and public opinion with regard to this country as we have in the last two years." Typically, people cite several things with regard to this. One was the Kyoto protocol, correct?
>
> **DASCHLE:** Correct.
>
> **SNOW:** You voted against that.
>
> **DASCHLE:** I did.
>
> **SNOW:** OK. The International Criminal Court, you voted against that.
>
> **DASCHLE:** That's correct.
>
> **SNOW:** And Iraq, where you voted with the president. So on all these key issues, the ones that the Europeans are constantly citing, you're on the same side as the White House.

This prompted our hero to voice his real complaint:

DASCHLE: Well, it's not necessarily the position in that legislative approach that I think is the concern. It's the attitude. It's the way that we have gone about foreign policy, especially, Tony, this unilateral approach to foreign policy, *dictating on a unilateral basis what the United States' position is going to be* and expecting, really, all these countries in a very autocratic or very authoritarian way to comply (emphasis added).

Yes, he really said that.

There we have it. The Democratic Party's then leader actually affirmed his belief that leaving it up to America to decide America's position is a "dictatorial approach [that] isn't going to work."

Left unanswered was precisely which nations should determine "what the United States' position is going to be." *Libya? The Maldives?* Of course! Europe.

The Holy Grail

U.S. ratification of Kyoto is the holy grail of Al Gore and other greens. Next time you end up at a cocktail party with one of these types, remember:

Kyoto would make us poorer. The *need* for Kyoto is based on faulty models. If the predictions are true, then Kyoto would be one-thirtieth of the needed fix. It is designed to fail, and is working as designed.

That is, Kyoto would be a prescription for addressing "global warming," but with

Kyoto is "the first component of an authentic global governance."

French president **Jacqes Chirac,** Opening remarks, Sixth Conference of Parties to Kyoto (COP-6), The Hague, November 2000

some not-so-minor caveats, such as: If it functioned as advertised and not as designed; if it didn't prove, as it has, to largely just send economic growth offshore; if Europe stopped gaming the system it created and began matching its rhetorical zeal for emissions reduction with action; and if the rest of the world stopped rejecting its scheme.

Finally, the world would have to somehow overcome the inescapable: climate has always changed and it always will; Man has always adapted, the wealthiest societies adapt the best; reducing global wealth—as Kyoto will—only ensures we will be even more vulnerable to the one certainty of the global warming debate: unpredictable and occasionally severe weather.

Europe is increasing emissions, not reducing them to meet its targets, and their elites are exposed as seeing the treaty as a Trojan horse for "global governance" that will "level the playing field."

Cattle That Rattle

The following is from the technical report of the European Environment Agency's 2006 "Greenhouse Gas Inventory":

Austria: Emissions of cattle dominate the trend. The reduction of dairy cows is partly counterbalanced by an increase in emissions per animal (because of the increasing gross energy intake, milk production and [nitrogen] excretion of [dairy] cattle since 1990).

CONCLUSION

President Dwight Eisenhower left office warning in his farewell address not just of a "military-industrial complex" but, in the next, less noticed paragraph, of a scientific-government complex. His prescience is, depressingly, becoming ever clearer. Even United States senators now feel comfortable threatening—in writing, no less— those who disagree with their views on Manmade global warming.

As the curtain descends on the remnants of scientific inquiry into and free speech about "environmental" and other such issues of controversy, we confront a circumstance in which a naturally driven climate is seized upon to cow a population with fear by governments seeking to expand their powers and businesses itching to profit from Man's gullibility. But it isn't over, yet.

The future does not have to be like the recent past. Simply opening the debate and holding it in the open air moves the ball from the alarmists' court—no time for questions, we must act now!—to the skeptics' court. Exercise your rights, indeed duties, of inquiry and speech, and demand that the future remain free, and full of energy.

$$\text{❋ ❋ ❋ ❋ ❋}$$

NOTES

Chapter 1
Green Is the New Red: The Anti-American, Anti-Capitalist, and Anti-Human Agenda of Today's Environmentalists

1. "The 12 biggest environmental pressure groups in the United States enjoy combined annual revenues of $1.9 billion, according to the latest Internal Revenue Service figures" wrote Hugo Gurdon in 2002, in "Grim Greens – and Greenbacks," November 22, 2002, at: http://www.cei.org/gencon/029,03290.cfm. "Only 725 of the United States' 20 million companies can boast such magnificent cash flow."

2. See, *e.g.,* Jonathan Adler, "Environmentalism at a Crossroads: Green Activism in America," Capital Research Center, January 1997.

3. See the list of corporations kowtowing to radical group Rainforest Action Network compiled by the Capital Research Center at http://www.capitalresearch.org/search/orgdisplay.asp?Org=RAN100, and its report, "Funding Liberalism With Blue-Chip Profits: Fortune 100 Foundations Back Leftist Causes," David Hogberg and Sarah Haney, August 2006, found at: http://www.capitalresearch.org/pubs/pdf/FW0806.pdf. CRC also details most pressure group contributor bases, and corporate and foundation giving

hen's excellent treatment of the phenomenon in "The iness: Environmentalist Groups Toe Funders' Line," Cap-r, July 2006, excerpt found at: http://www.capitalresearch. 706.pdf.

5. See, *e.g.*, C. Chumley, "United Nations Wages War on Capitalism," *Capitalism* magazine, August 15, 2004, found at: http://www.capmag.com/article.asp?ID=3843.

6. See generally the work of the Population Research Institute on the UN Population Research Fund at: http://www.pop.org.

7. See discussion on the United Nations Framework Convention on Climate Change, and its Intergovernmental Panel on Climate Change, discussed in Parts II and III, *infra*.

8. Opening remarks, 6th Conference of the Parties to the UNFCCC, The Hague, November 2000.

9. See some of Strong's finest quips at: http://www.brainyquote.com/quotes/authors/m/maurice_strong.html.

10. As a leading force in the strongly anti-NATO German Green Party, Kelly claimed to be "very tolerant" of communists, fearful not that Green Reds would be sympathetic to the Soviets, but too cozy with their own Social Democrats. These colleagues included an avowed Marxist-Leninist elected to a leadership position, Rainer Trampert, about whom the *New York Times* wrote at the time of his ascendance, "Trampert's quasi-Communist past does not trouble many of the Green faithful, who seem allergic to the deep anti-Communism of their elders." "Germany: For Greens it's Make Waves, Not War," James M. Markham, October 3, 1982, found in *Green World Press Review* at: http://ecquologia.it/sito/pag706.map?action=single&field.joined.id=49752&field.joined.singleid=52839.

11. http://www.eia.doe.gov/emeu/cabs/chinaenv.html.

12. This refers to the 1962 Rachel Carson tome generally credited with energizing the green movement.

13. See, *e.g.*, National Center for Public Policy Research, "League of Conservation Voters Scorecard Ignores Important Environmental Votes, But Includes Abortion and Campaign Finance Reform," February 28, 2002, found at: http://www.nationalcenter.org/TSR22802.html. This referred to the pressure group League of Conservation Voters assessing candidates' stances on the "Mexico City" policy which prohibits U.S. taxpayer-funded "family-planning" "foreign aid" from going to groups involved in the abortion industry; see *e.g.*, http://www.lcv.org/images/client/pdfs/2001_Scorecard_Final.pdf or http://www.lcv.org/scorecard/index.asp.

14. J. Taylor, "Greeniacs in Jo-burg: The U.N.'s latest 'Earth Summit,'" *National Review*, September 16, 2003.

15. See, *e.g.*, the work and materials of the Center for Private Conservation, at http://prfamerica.org/CEI-CPC.html.

16. As one Canadian diplomat-type was overheard saying during the early Clinton years, the new guys didn't get it. Every tragicomedy has its roles. At the UN, when one country would bang their shoes on the dais and demand green this and that, that cued the U.S. to say no. Instead, with the Clinton-Gore ideologues, "these guys agree to anything."

17. For a discussion of the reality that inequality does not in fact kill, see "Inequality and Mortality: Long-Run Evidence from a Panel of Countries", By Andrew Leigh and Christopher Jencks, Working Paper Number:RWP06-032, Submitted: 07/28/2006, John F. Kennedy School of Government Faculty Working Paper Series, found at:
http://ksgnotes1.harvard.edu/research/wpaper.NSF/rwp/RWP06-032.

18. I observed these very claims in Buenos Aires, December 2004, during the Inuits' "Right to be Cold" event. See, *e.g.*, M. Morano, "Warm homes causing Arctic ice melt, Eskimo charges," *CNSNews*, December 9, 2005, found at: http://www.cnsnews.com/ViewNation.asp?Page=%5CNation%5Carchive%5C200512%5CNAT20051209b.html.

19. "Is 'Conservative Environmentalist' an Oxymoron? How to End Environmental Policy Gridlock," August 2, 2005, at: http://www.aei.org/publications/filter.all,pubID.22934/pub_detail.asp.

20. In Wyoming Sawgrass v. US Forest Service, 383 F.3d 1241 C.A.10 (WY, 2004)(declined by the U.S. Supreme court in 2005) plaintiffs argued that a Forest Service historic monument designation was illegally based on an area's sacred nature to an Indian. The 10th Circuit gives the claims fairly respectful and detailed treatment. In the Minnesota deep ecology case, Associated Contract Loggers, Inc. v. U.S. Forest Service, 84 F.Supp.2d 1029 (D.Minn. 2000)(upheld by the 8th Circuit in 2001 and the Supreme court rejected the petition for *certiorari* in 2002), the court did view the claims as crazy.

21. Accessible via http://www.michaelcrichton.net/speeches/index.html.

22. Nelson's discussion, originally written in May 1996, in "Competitive Enterprise Institute Comments, OMB Draft Report to Congress on the Costs and Benefits of Federal Regulation", May 5, 2003, 129-152, found at:
http://www.cei.org/pdf/3515.pdf.

23. Note the utter negligence of policymakers also feigning concern about private forces—"gouging" or collusion—possibly producing high gas prices.

Congress holds enormous liability from the past decades of meddling in energy markets. In the 1970s price controls produced the inevitable result of gas lines. Literally thousands of inane, sometimes contradictory and always infuriating rules burden the U.S. refining industry such that it has insufficient refining and pipeline capacity; along with daunting barriers to entry these have borne no new refineries built in the U.S. in nearly 30 years. Add to this the ban on offshore oil and gas exploration applying to 97% of possible fields and EPA's and myriad state "boutique" fuel requirements, and who can be surprised by the price spikes of 2006 caused by severe weather (hitting an area where we have been forced to concentrate production, refining and import capacity) and Middle East uncertainties? Congress should be embarrassed, and ashamed. Instead, its Members point fingers devoid of any comprehension of the industry or the laws applying to it.

24. This absurd hierarchy has been advanced by personages ranging from Britain's chief science advisor Sir David King to former president Bill Clinton.

25. Nuclear power's only greenhouse emission is water vapor, but the relative contribution to is even more absurdly minimal than Man's contribution of CO_2 that no policy discussions include water vapor as a targeted "pollutant"; possibly because such an absurdity would give away the game.

26. Hayward, "Is 'Conservative Environmentalist' an Oxymoron?"

27. Bob Holmes, "Imagine Earth Without People," *New Scientist*, October 12, 2006.

28. See, *e.g.*, Lovelock's books *The Gaia Theory*, *The Ages of Gaia* and the autobiography *Homage to GAIA*. For a not totally unaccepting critique of this theory from a Christian perspective, see "The Gaia Hypothesis: Implications for a Christian Political Theology of the Environment," Stephen B. Scharper," http://www.crosscurrents.org/Gaia.htm.

Chapter 2

The Authoritarian Impulse: Environmentalists Want to Run Your Life

1. F. Hayek, *The Road to Serfdom* (Chicago, IL: The University of Chicago Press, 1994), xv-xvi.

2. Simon Jenkins, *The Sunday Times*, May 28, 2006.

3. M. McCarthy, "Climate change should be taken out of politics to allow radical remedies," *The Independent* (UK), July 13, 2006.

4. In fact the Department of Energy is pursuing such a program with Craig Venter, who previously decoded the human genome. See: http://doegenomestolife.org/news/111303abraham.shtml.

5. Ecolex, a website run in part by the UN Environment Programme, made this claim. Regardless, the ICC purported to provide a forum for prosecuting referrals for the war crime of intentionally and disproportionately harming the environment, if in language that is far less promiscuous than the promoters' original intent of the authority to prosecute "serious threats to the environment."

6. See text at: http://conventions.coe.int/Treaty/en/Treaties/Html/172.htm; the treaty, designed to "take effective measures to ensure that the perpetrators of environmental violations do not escape prosecution and punishment" (http://www.ecolex.org/en/treaties/treaties_full_display.php?docnr=3215&language=en) has eleven signatories.

7. Remarkably, the U.S. EPA co-sponsored an event at the 2002 Johannesburg World Environment Summit, see: http://www.un.org/events/wssd/press-conf/020827conf1.htm, follow up events to which affirm the participants' intent of fostering an international bar of green plaintiffs including to pursue novel ways of imposing their agenda through the courts. See also: http://www.iucn.org/themes/law/pdfdocuments/Pmb%20Env%20 Law%20Conf%20Prog_20%20June%202002.pdf.

8. See, *e.g.*, Friends of the Earth statement at: http://www.foe.org/new/ news17.html.

9. This is one of the more infuriating examples of green rhetoric and sloppy journalism combining. In fact, Kyoto Article 18 quite plainly states: "Any procedures and mechanisms under this Article entailing binding consequences shall be adopted by means of an amendment to this Protocol." This requires not only agreement on an amendment, but then approval by the parties, according to Kyoto's procedures. This has not occurred.

10. For example, deathless prose will no doubt ultimately soon spill forth claiming legitimacy for "global warming" damage suits under the theories of enterprise liability and the attorney general as public interest guardian. These are not (yet) serious propositions largely for the reasons touched on in this brief treatment.

11. For example, the Inuits' claim against the United States of threatening their "Right to be Cold," referenced elsewhere, was taken not to a court of law but a non-binding forum, an "Inter-American Commission for Human Rights," to which the U.S. is not even party. See petition and acknowledgement of same at: http://www.ciel.org/Publications/COP10_Handout_ EJCIEL.pdf#search=%22inuit%20complaint%20Inter-American%20Commission%20on%20Human%20Rights%20not%20party%22.

12. For a nice left-wing perspective on this sad affair, see Brendan O'Neill, "Global warming: the chilling effect on free speech, The demonisation of 'climate change denial' is an affront to open and rational debate," October 6, 2006, found at http://www.spiked-online.com/index.php?/site/article/1782/.

13. See: http://www.marklynas.org/wind/bloggin/296.html.

Chapter 3
The Sky Is Falling

1. *2000 EPA Annual Report: Performance Results*, Section II (Clean Air), found at http://epa.gov/ocfo/finstatement/2000ar/ar00_goal1.pdf.

2. This was the November 1996 proposed revision to the Clean Air Act's NAAQS; see, *e.g.*, discussion "Costs and Benefits of the PM Standard" in "Can No One Stop the EPA?", A. Antonelli, Heritage Foundation, July 8, 1997, found at http://www.heritage.org/Research/PoliticalPhilosophy/BG1129.cfm.

3. Eleventh Edition, April 2006, S. Hayward, American Enterprise Institute and Pacific Research Institute, found at: www.aei.org/docLib/20060413_2006Index.pdf.

4. *The Skeptical Environmentalist: Measuring the Real State of the World*, Cambridge University Press, 2001.

5. Given, however, that I had been informed of the fun, it was summer and I have two large breed dogs whose accumulated week's product can be quite pungent, it is fair to say Greenpeace's DoR might have considered delegating.

6. J. Schwartz, "No Smog for the Fear Factory," May 3, 2006, found at: http://www.tcsdaily.com/article.aspx?id=050306F.

7. Id., citations omitted.

8. Id.

9. See, *e.g.*, "Comment: A review and critique of the EPA's rationale for a fine particle standard," Suresh H. Moolgavkar, Regulatory Toxicology and Pharmacology 42 (2005) 123–144, March 2005. The standard cited above is that required to be met for ratcheting down the Clean Air Act's omnipotent and potentially economically crushing National Ambient Air Quality Standards ("NAAQS"). The law does not permit EPA to consider costs or otherwise economic impact, by the way, leaving no limit to what they may require and courts relatively powerless due to the doctrine of deferring to "agency expertise," vested though that expertise may be in ensuring alarm.

10. Greens would say that "it's not just one study. It's thousands of studies all showing that air pollution is killing thousands or causing them to get asthma or putting them in the hospital." And they would be right; there *are* thousands of epidemiological studies that purport to show that air pollution causes all manner of harm. But implementing an invalid technique thousands of times does nothing to increase its validity. See, *e.g.*, J. Schwartz, "Comments on EPA's Proposed Rule, National Ambient Air Quality Standards for Particulate Matter," Docket ID: EPA-HQ-OAR-2001-0017, April 17, 2006 found at http://www.joelschwartz.com/pdfs/Schwartz_PM25_NAAQS_041706.pdf.

11. J. Shields, "Greenpeace's fill-in-the-blank public relations meltdown," *Philadelphia Inquirer*, May 29, 2006, found at http://www.philly.com/mld/philly/news/14691089.htm.

12. Found at http://www.johnlocke.org/lockerroom/lockerroom.html?id=8090.

13. For a detailed treatment, see Marlo Lewis, "Judicial Activism in Overdrive: Massachusetts, et al., v. EPA," *Mealey's Pollution Liability Report*, August 2006, found at http://www.cei.org/pdf/5492.pdf.

14. Sound far-fetched? A former girlfriend related her experience with a news director at a major television network's flagship station in New York City who called the team, of which she was a part, together in the heat of the 1992 campaign to exhort them of their "journalistic duty to see that Bill Clinton is elected president!" What's a little lying about the environment in comparison?

15. http://org.eea.europa.eu/news/Ann1147868675/.

16. Brown played the same game year after year on total ocean fish production, which possibly has indeed capped out though ultimately attrib-

utable to myriad problems unrelated to Brown's thesis (for example, non-ownership of ocean fisheries, the staggering waste of perfectly good fish by-catch which is simply dumped overboard because the species are unfamiliar to the world's fish-eaters or have user-unfriendly names [remember the Toothfish switching its name to Chilean Sea Bass, and chefs couldn't get enough of it]). Minor changes—nearly all of them bureaucratic, could forestall the one time this blind pig found an acorn, for example by stiffing the milk-protein lobby to alter government health regulations and allow by-catch to simply be ground up whole, sterilized and bleached and then sold or given away as very inexpensive fish-protein powder to the world's malnourished and starving. Such solutions are far too sensible to ever be accepted, however, and we would no doubt be quite surprised at the doomsayers who find their way into that debate but to block such reform.

17. See "*Vital Signs 2006-2007*," available at: http://www.worldwatch.org/node/4344.

18. See, *e.g.*, "Debate on Climate Shifts to Issue of Irreparable Change: Some Experts on Global Warming Foresee 'Tipping Point' When It Is Too Late to Act," J. Eilperin, *Washington Post*, January 29, 2006.

19. See, *e.g.*, "It's too late to stop climate change: Interview with Hermann Ott," *Der Spiegel*, February 18, 2005, found at: http://service.spiegel.de/cache/international/0,1518,342431,00.html.

20. Thomas Wigley, "The Kyoto Protocol: CO2, CH4, and Climate Implications," *Geophysical Research Letter* 25 (1998): 2285-88.

21. See, *e.g.*, "This was hottest summer since 1936, report says," *USA Today*, September 15, 2006, found at http://www.usatoday.com/weather/climate/2006-09-13-hottest-summer_x.htm. See NOAA release and data at: http://www.noaanews.noaa.gov/stories2006/s2700.htm.

22. See, *e.g.*, David Malakoff, "Thirty Kyotos Needed to Control Global Warming," *Science*, 278, no.2, December 19, 1997, 2048.

23. To be precise, under the Marrakech amendments of 2001, Kyoto implicitly classifies generating electricity through nuclear power as yet one more threat greater than the supposed imminent, or already present, Man-made climate change. It does so by excluding nuclear as a permissible method of satisfying the treaty's CO2 reductions through the aid-to-the-poor-world CDM.

Nukes are the sole known "GHG-free" technology capable of providing our energy needs, emitting only water vapor. In fairness, water vapor is far and away the most prolific GHG, but the relevant quantities make this a non-issue.

24. http://www.wrm.org.uy/plantations/information/danger.html.

25. http://landscaping.about.com/cs/lazylandscaping/g/monoculture.htm.

26. *Nature*, January 12, 2006, vol. 439, 187. See more at: http://www.physorg.com/news9792.html.

27. Methane's "global warming potential" compared to CO_2 varies over the time-span chosen from 20 when considered over a century to much higher over, say 20 years (GWP = 60).

28. Z. Merali, "The lungs of the planet are belching methane," NewScientist.com, January 12, 2006.

29. Id.

Chapter 4
Global Warming 101

1. By the way, Man is blamed under the Kyoto Protocol for animal flatulence and, *e.g.*, rice paddy gases because we eat the product of both of these methane sources. The tonnage is calculated, and blame is assessed. This is the basis for the claim that agricultural operations will have to be shipped offshore under a Kyoto regime.

2. See, *e.g.*, global GHG budgets at: http://www.grida.no/climate/ipcc_tar/wg1/097.htm#tab31, which is not precisely the same thing as Earth's own production due to the natural uptake of carbon via the carbon cycle (*i.e.*, nature breathes out, plants breathe in).

3. As of this writing, this is a matter of rhetoric though not law, with the U.S. Supreme Court taking up the to-date failed action of *Commonwealth of Massachusetts v. U.S. Environmental Protection Agency*, in which greens seek to require EPA to treat CO_2 as if it were a pollutant.

4. To convert a Celsius temperature into Fahrenheit, multiply by 1.8 and add 32.

5. See science writer Ron Bailey's blog on same, at http://www.reason.com/rb/rb092206.shtml.

6. http://64.233.179.104/search?q=cache:ctDw6sczNv0J:www.senate.gov/~commerce.

7. See Mendelsohn submission to the written record, Senate Committee on Commerce, Science and Transportation, July 12, 2000, found at: http://commerce.senate.gov/hearings/0718men.pdf.

8. See, *e.g.*, "Katrina Should be a Lesson to US on Global Warming," *Speigel* Online, August 30, 2005, found at http://service.spiegel.de/cache/international/0,1518,372179,00.html. Of course, it is possible they were just imitating Hugo Chavez . . . and Robert F. Kennedy, Jr., see http://www.huffingtonpost.com/robert-f-kennedy-jr/for-they-that-sow-the-_b_6396.html. Even the breathless *New York Times* had to call this offense out as unsupportable, see "Storms Vary With Cycles, Experts Say," Kenneth Chang, August 30, 2005. For NOAA data on how unserious these claims are, see http://www.nhc.noaa.gov/pastdec.shtml, charted at http://eurota.blogspot.com/2005/08/eu-environmentalism-score-another-one.html.

9. See http://www.iea.org/textbase/press/pressdetail.asp?PRESS_REL_ID=163.

10. Peter Huber, "Hard Green: Saving the Environment from the Environmentalists," January 1, 2000. http://www.marshall.org/article.php?id=174.

Chapter 5
The "Consensus" Lie

1. http://www.john-daly.com.

2. http://people.aapt.net.au/~johunter/greenhou/home.html.

3. "The press gets it wrong: our report doesn't support the Kyoto treaty," *Wall Street Journal*, June 11, 2001.

4. Ibd.

5. Found at http://lwf.ncdc.noaa.gov/oa/aasc/AASC-Policy-Statement-on-Climate.htm.

6. T. Corcoran, "Climate consensus and the end of science," *Financial Post*, June 12, 2006.

7. See "Open Kyoto to Debate," *National Post*, found at: http://www.canada.com/nationalpost/news/story.html?id=3711460e-bd5a-475d-a6be-4db87559d605.

8. http://www.ucar.edu/news/releases/2005/ammann.shtml.

9. Wahl and Ammann finally found a home in February 2006, after much tribulation, in the journal *Climate Change*, published by none other than Stephen Schneider of "scary stories" fame.

10. Though green NGO activists are most likely to fall back on this convincing evidence of having either no case or no ability to argue it, they are hardly alone. In December 1997, ABC News *Nightline* anchor Ted Koppel replied to a global warming skeptic with "I was just going to make the observation that there are still some people who believe in the Flat Earth Society, too, but that doesn't mean they're right." See, *e.g.*, http://www.mediaresearch.org/cyberalerts/1997/cyb19971210.asp#3.

11. Gore refers to a several hundred word essay, or by its tone one might say op-ed, "Beyond the Ivory Tower: The Scientific Consensus on Climate Change," Naomi Oreskes, *Science* December 3, 2004: Vol. 306. no. 5702, p. 1686, DOI: 10.1126/science.1103618.

12. There are 11,149 documents listed on the ISI data base on "climate change" for the years 1993-2003, see http://portal.isiknowledge.com/portal.cgi.

13. Ph.D., Geology, now teaching in the Department of History and Science Studies Program, University of California at San Diego.

14. See "Naomi Oreskes Asks the Wrong Climate Question," D. Wojick, at http://maize-energy.blogspot.com/2006/07/naomi-oreskes-asks-wrong-climate.html.

15. Dr. B.J. Peiser, "The Dangers of Consensus Science", *National Post*, 17 May 2005, found at http://www.canada.com/national/nationalpost/news/story.html?id=b93c1368-27b7-4f55-a60e-5b5d1b1ff38b. Peiser was severely assailed by the alarmists when he made his attempts at replication public, including for the audacity of checking the two hundred additional abstracts in the same data bank but not mentioned by Oreskes.

16. Peiser has posted the abstracts allegedly analyzed by Oreskes on his website, at http://www.staff.livjm. ac.uk/spsbpeis/Oreskes-abstracts.htm. He notes in an email that the vast majority of abstracts doesn't even mention anthropogenic climate change, let alone support the consensus Oreskes claims that 75% either explicitly or implicitly endorse."

17. "Letters. CORRECTIONS AND CLARIFICATIONS: Essays: 'The scientific consensus on climate change' by N. Oreskes (3 Dec. 2004, p. 1686). The final sentence of the fifth paragraph should read 'That hypothesis was tested by analyzing 928 abstracts, published in refereed scientific journals between 1993 and 2003, and listed in the ISI database with the keywords "global climate change" (9).' The keywords used were 'global climate

change,' not 'climate change.'" *Science* 21 January 2005: Vol. 307. no. 5708, 355, DOI: 10.1126/science.307.5708.355.

18. See, "Failed Defense of Science Magazine Global Warming Study Fails to Address Critiques," Release, U.S. Senate Committee on Environment and Public Works, July 24, 2006, found at: http://www.epw.senate.gov/pressitem.cfm?party=rep&id=259323.

19. N. Oreskes, "Undeniable Global Warming," *Washington Post*, Sunday, December 26, 2004.

20. See, *e.g.*, this author's blog post *"H for Vendetta"*, CEI Open Market, March 22, 2006, at http://www.ceiopenmarket.org/openmarket/2006/03/h_for_vendetta.html.

21. According to its website, the Cosmos Club of Washington, DC was founded for "the advancement of its members in science, literature, and art," and "their mutual improvement by social intercourse." http://www.cosmos-club.org/main.html.

22. S. Fred Singer, C. Starr, and R. Revelle, "What To Do About Greenhouse Warming: Look Before You Leap," *Cosmos* 1 (1991): 28–33. Singer reveals the unsavory story in all its lurid details in a chapter of *Politicizing Science*, titled "The Revelle-Gore Story: Attempted Political Suppression of Science: The Alchemy of Policymaking," 283-297, Hoover Institution Press (Stanford, CA), and George C. Marshall Institute (Washington, DC), 2003, Michael Gough, Ed.

23. Easterbrook, "Green Cassandras," *New Republic*, July 6, 1992, pp. 23–25.

24. Singer, "The Revelle-Gore Story: Attempted Political Suppression of Science."

25. Id., citations omitted.

26. See also *e.g.*, "Global Warming Libel Suit Settled: Gore Tries To Suppress Scientists Views Again," found at http://www.suanews.com/articles/1994/globalwarminglibelsuitsettled.htm.

27. See, *e.g.*, R. Bailey, "Political Science," published in *Reason* magazine December 1993, found at: http://www.sepp.org/controv/happer.html.

28. Id.

29. Id.

30. "Comments on the Third U.S. Climate Action Report (CAR) by The Environmental Protection Agency," Michael R. Fox Ph.D. December 10, 2001, found at: http://yosemite.epa.gov/oar/globalwarming.nsf/UniqueKeyLookup/SHSU5BULL5/$File/fox.pdf begin on p. 3 of 5.

31. Cooney is a lawyer, not a scientist, though one who spent a decade working on the relevant issues sufficiently and with sufficient diligence that he was singled out by the Gore/Big Green attack machine. In the name of disclosure I am able to attest to Cooney's character arising from knowing him professionally from meetings we both attended during his days at an oil industry trade association, and as often as not in disagreement but sufficiently to make the assessment that he is a sober, honest professional.

32. These charges appeared in numerous sources are fairly well summed up in the NOW transcript found at http://www.cei.org/gencon/023,04717.cfm.

33. *The New Yorker*, "Comment: Ozone Man", David Remnick, April 24, 2006.

34. Please view the interview, after suffering through a long alarmist lead-in, available by scrolling down at http://www.cei.org/dyn/view_Expert.cfm?Expert=148.

35. Found at http://www.juliansimon.com/reply-critics.html.

36. See, *e.g.*, a chronicle of some green reactions reported, sometimes parroted, by media outlets (and some praise), in "Easterbrook's Moment: Reporters' Hour of Reaction", Environmental Health Center, *Environment Writer*, Vol. 7 No. 4, July 1995, at http://www.nsc.org/EHC/ew/issues/ew95jul.htm#easterb.

37. As quoted at Id.

38. See, *e.g.,* discussion in P. Michaels and Tereza Urbanova, "The Infection of Science by Public Choice: Steven Schneider vs. Bjorn Lomborg and *The Skeptical Environmentalist*", Competitive Enterprise Institute, December 2003, found at https://www.cei.org/pdf/3786.pdf.

39. See, *e.g.,* announcement with links to relevant documents, by Lomborg's Environmental Assessment Institute, an office established by the Danish government for him to continue his work, December 17, 2003, found at http://www.imv.dk/Default.aspx?ID=233.

40. http://www.climatesciencewatch.org/index.php/csw/details/earth-beat-interview/. Piltz is marginally familiar to me, as the lead player in a film project the crew behind which, in seeking an interview with me, proceeded to describe him as a scientist whose work they are chronicling. The same crew also followed me around for some time as their apparent villain, even leaving behind their emailed instructions in Portland, Oregon, despite the abundantly clear instructions "DO NOT LEAVE THIS LYING

AROUND!" See, *e.g.*, http://www.campusreportonline.net/main/ articles.php?id=416. At the December 2005 Montreal talks on Kyoto one member of this "Melting Planet" film team represented that Piltz was their "scientist"; I questioned this representation given Piltz's scientific pedigree, which he shrugged off as immaterial.

41. Specifically, the AGU claimed "There currently is insufficient skill in empirical predictions of the number and intensity of storms in the forthcoming hurricane season. Predictions by statistical methods that are widely distributed also show little skill, being more often wrong than right." American Geophysical Union, "Hurricanes and the U.S. Gulf Coast: Science and Sustainable Rebuilding," June 2006, found at: http://www.agu.org/report/hurricanes/.

Chapter 6
Getting Hot in Here?

1. See, Lubos Motl blog, The Reference Frame, September 25, 2006, found at: http://motls.blogspot.com/2006/09/southern-hemisphere-ignores-global.html.

2. For discussion and plots from the data and links to the research, see Steven McIntyre's blog, "New Satellite Data," September 24, 2006, at: http://www.climateaudit.org/?p=831.

3. http://www.ncdc.noaa.gov/oa/climate/research/cag3/na.html.

4. Thanks to Allan MacRae for his analysis.

5. See, *e.g.*, D. Deming, "Global Warming, the Politicization of Science, and Michael Crichton's 'State of Fear'", published in the June, 2005, issue of the *Journal of Scientific Exploration*, v.19, no.2, also found at: http://www.sepp.org/Archive/NewSEPP/StateFear-Deming.htm.

6. M. Mann, *et al.*, "Global-scale temperature patterns and climate forcing over the past six centuries", *Nature* Vol. 392, April 23, 1998 779-787, found at ftp://holocene.evsc.virginia.edu/pub/mann/mbh98.pdf.

7. Mann again revised the Hockey Stick in 2005 to cover 2,000 years, found at http://www.ncdc.noaa.gov/paleo/pubs/mann2003b/mann2003b.html.

8. http://www.grida.no/climate/ipcc_tar/wg1/fig2-21.htm and http://www.grida.no/climate/ipcc_tar/wg1/fig2-20.htm.

9. See GPUSA statement on Global Climate Change, at http://www.green-party.org/climate.php (last visited, September 28, 2006).

10. This is of a piece with the *Day After Tomorrow* scenario of shutdown of thermohaline circulation (THC). See, *e.g.*, "OUTLOOK If the increase in atmospheric greenhouse gas concentration leads to a collapse of the Atlantic THC, the result will not be global cooling. However, there might be regional cooling over and around the North Atlantic, relative to a hypothetical global-warming scenario with unchanged THC. By itself, this reduced warming might not be detrimental." "Abrupt Climate Change: Inevitable Surprises", The National Academies Press, release, comment and advance text of Chapter 4, "Global Warming as a Possible Trigger for Abrupt Climate Change," at "OCR for Page 115," found at: http://darwin.nap.edu/books/0309074347/html/107.html (2002).

11. The argument goes that "regional cooling is to be expected," caused by upwind industrial facilities leaving aerosols to cause this effect. See, *e.g,*, the cleverly worded "The warming is not uniform everywhere and there are in fact regions where one might actually expect regional cooling." Ask the Experts: Environment, ScientificAmerican.com, September 28, 2006.

12. "When the government says, 'The 20th century has been the warmest globally in the past 1,000 years,' it reports an untruth. Even the United Nations IPCC didn't go that far. It said there isn't 1,000 years of data for the southern hemisphere, so there's no way of knowing the 1,000-year history of global temperatures. What the IPCC did say is that it is 'very likely' the 1990s was the warmest since 1861. It also said it was 'likely' (as opposed to 'very likely') that 'the increase in surface temperature over the 20th century for the northern hemisphere' was greater than any century in the past 1,000 years." T. Corcoran, "See the Truth on Climate History", *National Post* (Canada), July 12 2006.

13. "Surface Temperature Reconstructions for the Last 2,000 Years" (2006)

14. Tree rings are generally thicker—meaning trees thrive—when temperatures and CO2 levels are higher.

15. "Climate Change Impacts on the United States, The Potential Consequences of Climate Variability and Change", National Assessment Synthesis Team, US Global Change Research Program, November 2000, found at http://www.usgcrp.gov/usgcrp/Library/nationalassessment/2IntroB.pdf.

16. You can view the results at http://landshape.org/enm/?p=15, a website allowing you to run the experiment yourself, at http://landshape.org/enm/?p=48. The precise results are rarely put into plain English for a ready cite here. For a detailed but accessible explanation of how and why similar results will occur under Mann's program with most series of random numbers, see Steven McIntyre's clarifying discussion at http://www.climate2003.com/blog/pacific.htm.

17. McIntyre and McKitrick in *Energy and Environment*, 14,751-771, 2003, updated in "The M&M Critique of the MBH98 Northern Hemisphere Climate Index: Update and Implications," McIntyre, Stephen; McKitrick, Ross, Energy & Environment, Volume 16, Number 1, January 2005, pp. 69-100(32). See also generally, "The *M&M Project: Replication Analysis of the Mann, et al., Hockey Stick*," found at: http://www.uoguelph.ca/~rmckitri/research/trc.html.

18. See, *e.g.*, "Don't believe in the null hypothesis?" at http://seamonkey.ed.asu.edu/~alex/computer/sas/hypothesis.html.

19. See, *e.g.*, global-cooling-*cum*-warming aficionado Stephen Schneider, of "choice between being honest and being effective" fame, http://stephenschneider.stanford.edu/Climate/Climate_Science/CliSciFrameset.html?http://stephenschneider.stanford.edu/Climate/Climate_Science/Contrarians.html.

20. S. McIntyre, R. McKitrick, "Does the Hockey Stick Matter?", posted November 14, 2005, found at: http://sciencepolicy.colorado.edu/prometheus/archives/climate_change/000630does_the_hockey_stic.html.

21. Id.

22. R. McKitrick, "What is the 'Hockey Stick' Debate About?", paper delivered April 4, 2005 for the APEC Study Group as an Invited Special Presentation to the Conference "Managing Climate Change—Practicalities and Realities in a Post-Kyoto Future," Parliament House, Canberra Australia, April 4 2005 found at http://www.climatechangeissues.com/files/PDF/conf05mckitrick.pdf.

23. McKitrick, 5.

24. Id.

25. See http://www.climateaudit.org/index.php?p=127.

26. Jan Esper, Robert J.S. Wilson, David C. Frank, Anders Moberg, Heinz Wanner, Jurg Luterbacher, "Climate: Past Changes and Future Ranges," *Quaternary Science Reviews*, 24 (2005), 2164-2166.

27. In the interests of full disclosure, in September 2006 I filed an *amicus* brief on behalf of Chairman Barton before the U.S. Supreme Court, challenging EPA's "New Source Review" rule. This matter is unrelated to "climate" and I have had no dealings with the chairman on this issue.

28. Letters are available at http://energycommerce.house.gov/108/Letters/062305_Mann.pdf.

29. Letter is available at: http://www.realclimate.org/Boehlert_letter_to_Barton.pdf.

30. S. McIntyre and R. McKitrick, "Misled again: The Hockey Stick climate: History is flawed, and so is the process by which its author's claims have been adjudicated," *National Post* (Canada), July 12, 2006.

31. The NAS's contract arm which actually performs these activities, the National Research Council, is no stranger to curious panel compositions when it comes to "global warming". Their panel reviewing the very first IPCC Assessment, in fact, "was hardly promising. It had no members of the National Academy expert in climate. Indeed, it had only one scientist directly involved in climate science, Stephen Schneider, who is an ardent advocate. It also included three professional environmental advocates..." *Environmental Gore: A Constructive Response to Earth in the Balance*, Richard Lindzen's chapter "Global Warming: the Origin of Consensus,"130.

32. Available online at http://www.nap.edu/catalog/11676.html.

33. "Study: Earth 'likely' hottest in 2,000 years," CNN.com, found at: http://edition.cnn.com/2006/TECH/science/06/22/global.warming.ap/index.html.

34. B. Daley, "Report Backs Global Warming Claims," *Boston Globe*, June 22, 2006.

35. Glen Johnson, "Kerry Outlines Updated Energy Plan," June 26, 2006 (emphasis added).

36. G. Brumfiel, "Academy affirms hockey-stick graph," *Nature*, 441, 1032-1033 (29 June 2006) doi:10. 1038/4411032a, published online June 28 2006.

37. R. Lyons, "Climate Change: A Model Cock-Up," Spiked Online, April 20, 2006, found at: http://www.spiked-online.com/Printable/0000000CB027.htm.

38. "US National Assessment of the Potential Consequences of Climate Variability and Change: A detailed overview of the consequences of climate change and mechanisms for adaptation," found at: http://www.usgcrp.gov/usgcrp/nacc/default.htm.

39. This author sued the Clinton Administration over this document on behalf of Senator Jim Inhofe (R.-Okla.), Representatives JoAnn Emerson (R-Mo.) and Joseph Knollenberg (R-Mich.), the Competitive Enterprise Institute, and several others. With key questions outstanding including the plaintiffs' ability to gain "standing" to sue, the result was an agreement under which plaintiffs would withdraw their complaint and the government would put a disclaimer on the website that the document had not been subjected to federal rules requiring that data be objective and reproducible.

40. Paul K. Driessen, "National Assessment of Climate Change released," *Environment News*, August 1, 2000, The Heartland Institute.

41. Found at: thttp://www.cato.org/pubs/wtpapers/michaels0206.pdf#search=%22pat%20michaels%20karl% 20random%20numbers%22.

42. Michaels' testimony is found at http://energycommerce.house.gov/107/hearings/07252002Hearing676/ Michaels1146.htm; the transcript of the hearing, by the U.S. House of Representatives' Committee on Energy and Commerce, Subcommittee on Oversight and Investigations, July 25, 2002, is available at http://energycommerce.house.gov/107/hearings/07252002Hearing676/print.htm.

43. P. Michaels, "Review of 2001 U.S. Climate Action Report," at 6.

44. Regarding the Hadley quote, since CEI brought attention to this damning admission in some if its regulatory filings, the Hadley Center subsequently wordsmithed this acknowledgement to make some its obvious meaning less obvious, though without substantively altering it.

45. The pressure group "Union of Concerned Scientists" promotes such unsupportable, state-specific outcomes (at least honestly styled as "potential", if so speculative as to not reasonably warrant such speculation), and California adopted the National Assessment's findings to present a state-specific climate scenario to justify imposing their unique Kyoto-style law.]

46. See, *e.g.*, "Modellers deplore 'short-termism' on climate," *Nature*, 428, 593, April 8, 2004, found at: www.nature.com/news/2004/040405/full/428593a.html (subscription required), brief abstract available at http://www.scidev.net/News/index.cfm?fuseaction=readNews&itemid=131 9&language=1. See also "Climate models have no answer to burning questions," *Nature* 424,867, August 21, 2003, brief abstract and link (subscription

required) found at http://www.scidev.net/dossiers/index.cfm?
fuseaction=dossier. ReadItem&type=1&itemid=968&language=1&dossier=
4&CFID=2395233&CFTOKEN=74261043.

47. For a discussion of media misrepresentation see, William O'Keefe and
Jeff Keuter, "Climate Models: A Primer," 2004, George C. Marshall Institute,
found at http://www.marshall.org/pdf/materials/225.pdf.

48. Stainforth, D. et al., "Uncertainty in predictions of the climate
response to rising levels of greenhouse gases," *Nature*, 433, 403-406.

49. "Don't Worry, Be Happy," April 22, 2004, found at http://www.
tcsdaily.com/article.aspx?id=042204B.

Chapter 7
Melting Ice Caps, Angrier Hurricanes, and Other Lies about the Weather

1. See, *e.g.*, *Meltdown: The Predictable Distortion of Global Warming by
Politicians, Science and the Media*, Patrick J. Michaels, Washington DC,
Cato Institute, 2004, 95-96, citing Przybylak (2000).

2. "Last Stand of Our Wild Polar Bears: Silly to predict their demise;
Starling conclusion to say they will disappear within 25 years and surprise to many researchers," Mitchell Taylor, The Toronto Star, May 1,
2006.

3. Id., citing, dissecting and otherwise straightening out Barbraud published in *Nature* (2001).

4. Polyakov, I., Akasofu, S-I., Bhatt, U., Colony, R., Ikeda, M., Makshtas,
A., Swingley, C., Walsh, D. and Walsh, J. 2002. Trends and variations in
Arctic climate system. *EOS, Transactions, American Geophysical Union*
83: 547-548.

5. Found at http://www.acia.uaf.edu/.

6. J. Kay, "Gray whales thrive in the Arctic, for now: more calves being
born, but the effects of warming not clear" (NB: sure seems clear to me:
warming causes whales to calve), *San Francisco Chronicle*, June 28, 2006.

7. Zwally, H.J., Giovinetto, M.B., Li, J., Cornejo, H.G., Beckley, M.A.,
Brenner, A.C., Saba, J.L. and Yi, D. 2005. "Mass changes of the Greenland
and Antarctic ice sheets and shelves and contributions to sea-level rise:
1992-2002." *Journal of Glaciology* 51: 509-527.

8. Mörner, N.A. 2003. "Estimating Future Sea Level Changes from Past
Records," *Global and Planetary Change* 40: 49-54. Mörner by the way is an

acknowledged expert, creatures that Mörner will tell you are in very short supply on the IPCC. The IPCC's inherent deficiencies are discussed, *infra*.

9. Working Group 1, Summary for Policymakers, 4.

10. Id., 10.

11. See, *e.g.*, "Despite the long term warming trend seen around the globe, the oceans have cooled in the last three years, scientists announced today.... Researchers have not yet identified the cause of ocean cooling in the last three years but hope that further studies will clarify this anomaly." "Global warming takes a break," Sara Goudarzi, LiveScience, September 21, 2006.

12. Mudelsee, M., *et al.*, 2003. "No upward trends in the occurrence of extreme floods in central Europe." *Nature*, 425, 166-169.

13. http://presszoom.com/story_113097.html.

14. The Director of the World Climate Program for the WMO, Ken Davidson, replied to a questioner in Geneva in 2003, "You are correct that the scientific evidence (statistical and empirical) are [sic] not present to conclusively state that the number of events have [sic] increased. However, the number of extreme events that are being reported and are truly extreme events has increased both through the meteorological services and through the aid agencies as well as through the disaster reporting agencies and corporations. So, this could be because of improved monitoring and reporting," quoted at: http://www.john-daly.com/press/press-03b.htm.

15. Quoted in W.Williams, "The Politics of Hurricanes and Global Warming", October 3, 2006, found at http://www.capmag.com/article.asp?ID=4425.

16. www.nhc.noaa.gov/pastdec.shtml.

17. NOAA Technical Memorandum NWS TPC-4, Blake, *et al.*, found at http://www.nhc.noaa.gov/pdf/NWS-TPC-4.pdf.

18. "THE DEADLIEST, COSTLIEST, AND MOST INTENSE UNITED STATES TROPICAL CYCLONES FROM 1851 TO 2005," http://www.nhc.noaa.gov/Deadliest_Costliest.shtml.

19. See "Climate Change and Reinsurance, Part I", January 6, 2005, found at http://sciencepolicy.colorado.edu/prometheus/archives/climate_change/0003 11climate_change_and_r.html.

20. R. A. Pielke, Jr., C. Landsea, M. Mayfield, J. Laver, R. Pasch., "Global Warming and Hurricanes", *Bulletin of the American Meteorology Society*, November 2005, pgs. 1571-75, found at

http://sciencepolicy.colorado.edu/admin/publication_files/resource-1766-2005.36.pdf.

21. Klotzbach, P. J. (2006), "Trends in global tropical cyclone activity over the past twenty years (1986–2005)," *Geophys. Res. Lett.*, 33, L10805, doi:10.1029/2006GL025881.

22. C. Landsea, B. Harper, K. Hoarau, J. Knaff, "Can we detect trends in extreme tropical cyclones?", *Science* 28 July 2006 313: 452-454 [DOI: 10.1126/science.1128448], July 28, 2006.

23. See, *e.g.*, Associated Press, "Global warming's effect on hurricane strength disputed in new report," *Orlando Sun-Sentinel,* July 28, 2006, found at http://www.sun-sentinel.com/news/local/southflorida/sfl-0729globalwarming,0,3801546.story.

24. A. Naparstek, "Storm Tracker: a history of hurricanes in New York—including the day in 1893 that Hog Island disappeared for good", *New York* Magazine, September 12, 2006 found electronically at: http://www.newyorkmetro.com/nymetro/news/people/columns/intelli-gencer/12908/.

25. "Bizarre Weather Ravages Africans' Crops: Some See Link To World-wide Warming Trend," Michael Grunwald, *Washington Post*, January 7, 2003.

26. *Geophysical Research Letters*, Vol. 33, L10403, doi:10.1029/2006GL025711, published May 25, 2006.

27. http://www.grida.no/climate/ipcc_tar/wg1/092.htm.

28. See, *e.g.*, "Plants fighting back against African desert areas," Reuters, September 19, 2002, found at http://www.planetark.org/dailynewsstory.cfm/newsid/17813/story.htm.

29. A. Gore, *An Inconvenient Truth*, p. 106, citing the Millennial Ecosystem Assessment.

30. See "Climate Change and Reinsurance, Part I", January 6, 2005, found at http://sciencepolicy.colorado.edu/prometheus/archives/climate_change/0003 11climate_change_and_r.html.

31. Id., citing the MEA Chapter 16, "Regulation of Fires and Floods," Table 16.5, 447.

32. Emergency Disasters Database, found at http://www.em-dat.net/index.htm.

33. "Millennium Ecosystem Assessment," 447 (emphases added).

34. Id., 448.

Chapter 8

Media Mania

1. J. Schwartz, "Air Pollution and Health: Do Popular Portrayals Reflect the Scientific Evidence?," *Environmental Policy Outlook* No. 2, 2006 (citations omitted), found at: http://www.aei.org/publications/pubID.24313/pub_detail.asp.

2. Alaskan temperatures have tended to warm during the positive phases of the Pacific Decadal Oscillation (PDO). See, *e.g.,* S. Hare and N. Manua, http://tao.atmos.washington.edu/pdo/.

3. Chapter 1, "Where all is South," of Revkin's *The North Pole Was Here* as pre-released.

4. See "Alaska's Dragons," *Alaska Wildlife News*, August 2004, found at: http://www.wc.adfg.state.ak.us/index.cfm?adfg=wildlife_news.view_article&issue_id=17&articles_id=70.

5. Edmund Blair Bolles, "In the (Un)Frozen North," *New York Times*, August 23, 2000.

6. See http://www.realclimate.org/index.php?cat=26.

7. G. Will, "Let Cooler Heads Prevail: The Media Heat Up Over Global Warming," *Washington Post*, April 2, 2006.

8. CBS Public Eye blog, March 23, 2006, found at: http://www.cbsnews.com/blogs/2006/03/22/publiceye/entry1431768.shtml.

9. June 13, 2006, found at: http://abcnews.go.com/Technology/story?id=2067792&page=1.

10. Blakemore's remarks can be found at http://abcnews.go.com/US/print?id=2374968. See a rebuttal to ABC incorporating these and other sources, by the Senate Environment and Public Works Committee, found at http://ff.org/centers/csspp/library/co2weekly/20060907/20060907_01.html.

11. Inhofe speech available at http://epw.senate.gov/speechitem.cfm?party=rep&id=263759.

12. "America reacts to speech debunking media global warming alarmism," Senate EPW Committee, release, September 28, 2006, found at: http://www.epw.senate.gov/pressitem.cfm?id=264042&party=rep.

13. See, *e.g.*, http://www.nsf.gov/statistics/seind00/access/c8/c8s5.htm.

14. See, *e.g.*, M. Morano, "The Weather Channel Warms up to Climate Change Theory", CNSNews.com, April 14, 2005, found at:

http://www.cnsnews.com//ViewSpecialReports.asp?Page=/SpecialReports/archive/200504/SPE20050414a.html.

15. See, *e.g.,* M. Morano, "Scientist Alleging Censorship Bush helped Gore, Kerry," CNSNews.com, March 23, 2006, found at: http://www.cnsnews.com/ViewPolitics.asp?Page=%5CPolitics%5Carchive%5C2006 03%5CPOL20060323a.html.

16. "Greenpeace pictures as stunt—Global warming claim meaningless as glacier photos show 'natural changes in shape,'" Jo Knowsley, *Daily Mail on Sunday*, August 11, 2002, found at http://www.scientific-alliance.org/news_archives/climate/greenpeacestunt.htm.

17. See "The Economist and the Greenpeace Glacier Photo Stunt," at: http://ff.org/centers/csspp/library/co2weekly/20060920/20060920_17.html.

18. "Fire and Ice: Journalists have warned of climate change for 100 years, but can't decide weather we face an ice age or warming," BMI Special Report, 2006, found at http://www.businessandmedia.org/specialreports/2006/fireandice/fireandice.asp.

19. "Yelling Fire on a Hot Planet: Global warming has the feel of breaking news these days," Andrew Revkin, *New York Times*, April 23, 2006, found at: http://www.nytimes.com/2006/04/23/weekinreview/23revkin.html?ex=1303444800&en=a9aa22cef0624748&ei=5090&partner=rssuserland&emc=rss.

20. http://www.bbc.co.uk/sn/hottopics/climatechange/climatechaos.shtml.

21. http://www.bbc.co.uk/pressoffice/pressreleases/stories/2006/05_may/24/green.shtml.

22. "Grist Soapbox," August 24, 2006, found at: http://www.grist.org/comments/soapbox/2006/08/24/reporters/.

Chapter 9
The Big Money of Climate Alarmism

1. R. Novak, "Enron's Green Side," *Chicago Sun-Times*, January 17, 2002.

2. Readers may also want to read Paul Sperry's series on the same subject on www.WorldNetDaily.com.

3. "Teresa Heinz answers her critics," *Time* magazine, February 3, 2004, available at: http://www.time.com/time/covers/1101040209/nheinz_interview.html.

4. See, *e.g.,* P. Gigot, "Not So Hot: Bush's Treasury Secretary gets off to a bad start," *Wall Street Journal*, March 16, 2001, found at www.opinionjournal.com/columnists/pgigot/?id=85000712.

5. "BP's failure of execution, not strategy," Craig Smith, *Financial Times*, August 9, 2006.

6. J. Quelch, "How soft power is winning hearts, minds and influence," *Financial Times*, October 10, 2005.

7. Ibid.

8. See, *e.g.*, the Pew gang's October 18, 2006, call to action, found at: http://www.pewclimate.org/press_room/sub_press_room/2006_releases/pr_1018.cfm.

9. This is according to colleagues expert in these matters and supported in documents such as "Greenhouse Gas Emission Reduction Verification Audit for DuPont Canada Inc.'s Maitland [sic] Ontario Apidic [sic] Acid Plant, Final Report, February 20, 2002," at http://www.ghgregistries.ca/files/projects/prj_9860_644.pdf.

10. See http://www.cinergy.com/presentations/091505_analyst_meeting.ppt#19.

11. See November 2005 Climate Science Workshop of the United States Climate Change Science Program at http://www.climatescience.gov/workshop2005/generalinfo.htm.

12. Stowell's "flat-Earth" slur was either *ex temp* or intentionally not included in his prepared remarks when later posted, at http://www.climatescience.gov/workshop2005/presentations/Mon_plenary_Stowell.htm.

13. Kevin Morrison, "Morgan Stanley Makes $3bn Green Pledge," *Financial Times*, October 26, 2006.

Chapter 10

Al Gore's Inconvenient Ruse

1. http://grist.org/news/maindish/2006/05/09/roberts/index.html.

2. "In March 1987, I decided to run for president. . . . In the speech in which I declared my candidacy, I focused on global warming, ozone depletion and the ailing global environment and declared that these issues—along with nuclear arms control—would be the principal focus of my campaign." *Earth in the Balance*, 8.

3. See "In 'docu-ganda' films, balance is not the objective," Daniel B. Wood, June 2, 2006, found at: http://www.christiansciencemonitor.com/2006/0602/p01s02-ussc.html.

4 A. Gore, *An Inconvenient Truth*, Rodale Books, 2006.

5. See, *e.g.*, P.J. Klotzbach, "Trends in global tropical cyclone activity over the past twenty years (1986–2005)," GEOPHYSICAL RESEARCH LETTERS, VOL. 33, L10805, doi:10.1029/2006GL025881, 2006.

6. See discussion with links of CNN anchor Miles O'Brien referring to this movie in support of his climate alarmism in his September 2006 interview with Senator James Inhofe, at: http://thecablegame.blogspot.com/2006/10/its-only-movie-miles-its-not-real-even.html.

7. Even the alarmist publications such as *Nature* and *Science* magazine have published work acknowledging this. See, *e.g.*, Dahl-Jensen, D., Mosegaard, K., Gundestrup, N., Clow, G.D., Johnsen, S.J., Hansen, A.W. and Balling, N. 1998. "Past temperatures directly from the Greenland Ice Sheet." *Science* 282: 268–271.

8. The Holocene is the name given to the last ~10,000 years of the Earth's history, since the end of the last major glacial epoch, or what we know as an "ice age." The Holocene has exhibited comparatively small-scale climate shifts—notably the "Little Ice Age" between about 1200 and 1700 A.D. which was preceded by the Medieval Climate Optimum, or warming, from about 900 A.D. Generally, however, even among climate alarmists the Holocene is recognized as having been a relatively warm period in between ice ages.

9. This chart is also found on 66–67 of Gore's book, and it is charitable to describe his intimating that the relationship between CO_2 and temperature is linear as simply wildly misleading.

10. P. Stanway, "An Inconvenient Truth for Gore," *Edmonton Sun*, July 7, 2006.

11. See www.uah.edu/News/newsread.php?newsID=291.

12. B. Carter, "There IS a problem with global warming . . . it stopped in 1998,"*Daily Telegraph* (UK), April 9, 2006.

13. See, *e.g.*, "The Making of a Heatwave," National Weather Forecasting Office, found at http://www.srh.noaa.gov/abq/feature/heat_wave.htm.

14. A.V. Fedorov, *et al.*, "The Pliocene Paradox (Mechanisms for a Permanent El Niño)", *Science* 9 June 2006: Vol. 312. no. 5779, pp. 1485 – 1489, DOI: 10.1126/science.1122666, (emphasis added); summary available

online at http://www.sciencemag.org/cgi/content/short/312/5779/1485, chart available online at https://www6.miami.edu/media/2006-06-12-sciencemag-lookingwayback.pdf.

15. In Gore's book, he discusses a Peruvian glacier that recent research indicates probably disappeared a few thousand years ago.

16. "Inconvenient Truths Indeed," Dr. Robert C. Balling Jr., Tech Central Station, May 24, 2006.

17. It appears that primitive—there's that word again—lifestyles and methods of dealing with threats is behind the Kilimanjaro area's deforestation ("Poverty is often at the root of the forest fires, which are set by illegal honey gatherers who burn sticks of wood to protect themselves against aggressive African bees"); see *Der Speigel's* treatment of the issue found at the M&M "Climate Audit" blog, at http://www.climateaudit.org/?p=554.

18. "Glacier Upsala: Nuevo Fraude de Greenpeace," http://mitosyfraudes. 8k.com/Ingles3/UpsalaEng.html. Mitos y Fraudes is of course an activist website created to counter, *e.g.*, global warming alarmism, however it does link to and is a good resource for relevant scientific literature.

19. Ibid., emphases supplied in original.

20. http://www.geo.unizh.ch/wgms/.

21. "Solar modulation of Little Ice Age climate in the tropical Andes," June 1, 2006, *Proc. Natl. Acad. Sci. USA*, 10.1073/pnas.0603118103, found at http://faculty.eas.ualberta.ca/wolfe/eprints/Polissar_PNAS2006.pdf#search= %22Polissar%20andes%22.

22. Comment, posted on CCNet, see October 5, 2006 archive found at http://www.staff.livjm.ac.uk/spsbpeis/CCNet-2006.htm.

23. "Scientists Predict Solar Downturn, Global Warming," *New Scientist*, September 16, 2006, available at http://ff.org/centers/csspp/library/co2weekly/20060920/20060920_13.html.

24. Petr Chylek, *et al.*, "Greenland warming of 1920–1930 and 1995–2005," *Geophysical Research Letters*, 33, L11707, June 13, 2006.

25. Ibid.

26. Ibid.

27. See T. Harris, "Scientists respond to Gore's warnings of climate catastrophe," *Canada Free Press*, June 12, 2006, found at http://www.canadafreepress.com/2006/harris061206.htm.

28. See discussion in P. Michaels, *Meltdown: The Predictable Distortion of Global Warming,* 51–54.

29. See the British Antarctic Survey's take found at http://www.antarctica.ac.uk/About_Antarctica/FAQs/ faq_02.html, including "Continent wide there is little evidence for increased melting of the land-based ice sheet. This is because around 99% of Antarctica is so cold and high that temperatures cannot rise above freezing. The average surface temperature of the Antarctic continent is around -37° C and at an 'average site' there is no melting on any day of the year. By contrast on the warmest coast of the Antarctic Peninsula average temperatures are around -5° C and there may be two months in the year with temperatures above freezing. Here the warming climate has led to enhanced melting, and some floating ice shelves have disintegrated because of this. . . . Melting is an essential part of the glaciological cycle that starts as the fall of snow over the continent and ends by melting. It is still difficult to tell whether more ice is currently melting from the Antarctic than arrives by snowfall. If it is, then the Antarctic ice sheet can only become smaller. And once the non-floating parts of the ice sheet start to diminish in volume so will sea level start to rise."

30. See, *e.g.*, British Antarctic Survey at http://www.antarctica.ac.uk/, maps available at http://www.photo.antarctica.ac.uk/external/guest/light-box/search/list/1.

31. Quoted in T. Harris, "Scientists respond to Gore's warnings of climate catastrophe," Canada Free Press, June 12, 2006, found at http://www.canadafreepress.com/2006/harris061206.htm.

32. The study to which Gore alludes [Velicogna, I., and J. Wahr, 2006, "Measurements of time-variable gravity show mass loss in Antarctica," *Sciencexpress*, March 2, 2006], measured ice mass changes in Antarctica based on only three years of data. But other studies (*e.g.*, Davis, C. H., *et al.*, 2005, "Snowfall-driven growth in East Antarctic ice sheet mitigates recent sea-level rise," *Science*, 308, 1898-1901, also Zwally *et al.* 2005, "Mass changes of the Greenland and Antarctic ice sheets and shelves and contributions to sea-level rise: 1992-2002," *Journal of Glaciology* 51: 509-527) estimate ice mass changes based on a decade of data.

33. See, *e.g.*, TAR "Summary for Policymakers," found at: http://www.ipcc.ch/pub/spm22-01.pdf.

34. Ibid. Although the alarmists' "bible" the IPCC (2001) found no acceleration in 20th century sea-level rise, a recent study—John Church and Neil White, "A 20th century acceleration in global sea-level rise," *Geophysical*

Research Letters, Vol. 33, L01602, doi:10.1029/2005GL024826, 2006—did find acceleration. However, even were that study to credibly overtake all prior work and the acceleration continues, there will be about 1 foot of sea level rise in the 21st century. Apocalypse not.

35. C. Wunsch, "Gulf Stream safe if wind blows and Earth turns," Letter, *Nature* magazine, April 8, 2004.

36. See, *e.g.,* R. Lindzen *et al,* "Does the Earth Have an Adaptive Infrared Iris?" *Bulletin of the American Meteorological Society*, March 2001, Vol. 82, No. 3, pp. 417–432, available at http://ams.allenpress.com/ amson-line/?request=get-abstract&issn=1520-0477&volume=082&issue=03&page=0417.

37. See, *e.g.,* Ron Bailey's testimony before the U.S. House Subcommittee on Energy and Mineral Resources, "Science and Public Policy," February 4, 2004, found at http://www.cei.org/pdf/3852.pdf.

38. Ibid.

39. *Washington Post*, Elizabeth Williamson, July 15, 2006.

40. T. Harris, "The Gods Are Laughing," June 7, 2006, *National Post*, found at http://www.canada.com/ nationalpost/financialpost/story.html?id=d0235a70-33f1-45b3-803b-829b1b3542ef&&p=4.

41. "Inconvenient Truths Indeed," Dr. Robert C. Balling Jr., May 24, 2006, found at http://www.tcsdaily.com/article.aspx?id=052406F.

42. B. Fagan, *The Little Ice Age: How Climate Made History (1300–1850)*, Basic Books, 2001.

43. Reiter, P. et al., "Global Warming and Malaria, A Call for Accuracy," *Lancet Infectious Diseases* 2004 Jun; 4(6):323–4.

44. These quotes are from Dr. Reiter's July 28, 1998, talk delivered in Washington, "Global Warming and Vector-Borne Disease: Is Warmer Sicker?" found at http://www.cei.org/gencon/014,01520.cfm.

Chapter 11
The Cost of the Alarmist Agenda

1. Hoffert et al., "Advanced Technology Paths to Global Climate Stability: Energy for a Greenhouse Planet," *Science*, November 1, 2002, Vol. 298. no. 5595, 981 – 987 DOI: 10.1126/science.1072357.

2. Ibid.

3. "All cost, no benefit," TechCentralStation.com, July 20, 2005, Marlo Lewis, found at http://www.cei.org/utils/printer.cfm?AID=5329 (citations omitted).

4. http://www.eia.doe.gov/emeu/international/gas1.jpg.

5. See, "Transport and environment: facing a dilemma," European Environment Agency Report No 3/2006, page 17. Found at: http://reports.eea.europa.eu/eea_report_2006_3/en/term_2005.pdf.

6. U.S. Energy Information Administration, "Impacts of the Kyoto Protocol on U.S. Energy Markets and Economic Activity," October 1998, found at http://www.eia.doe.gov/oiaf/kyoto/pdf/sroiaf9803.pdf.

7. "An Evaluation of Cap-and-Trade Programs for Reducing U.S. Carbon Emissions," June 2001, United States Congressional Budget Office. Chapter found at http://www.cbo.gov/showdoc.cfm?index=2876&sequence=2.

8. Pizer, William A. (1997). "Prices vs. Quantities Revisited: The Case of Climate Change." Resources for the Future Discussion Paper 98-02.

9. Energy Information Agency, "What Does the Kyoto Protocol Mean to U.S. Energy Markets and the Economy," U.S. Department of Energy, http://www.eia.doe.gov/oiaf/kyoto/market.html.

10. See, *e.g.,* Peter Huber and Mark Mills, "Got a computer? More power to you," *Wall Street Journal*, September 7, 2000. Available at: http://www.manhattan-institute.org/html/_wsj-got_a_comp.htm.

11. See, *e.g.,* "EU moots border tax to offset costs of climate action," Euractiv.com, October 11, 2006, found at http://www.euractiv.com/en/sustainability/eu-moots-border-tax-offset-costs-climate-action/article-158641.

12. See, *e.g.,* "Three Spanish companies closed down for violating Kyoto Protocol," *Spain Herald*, August 9, 2005.

13. See, *e.g.,* OpenEurope, "The high price of hot air: why the EU Emissions Trading Scheme is an environmental and economic failure," July 2006.

14. See, *e.g.,* Testimony of Dr. Margo Thorning, Senate Committee on Environment and Public Works, October 5, 2005, found at: http://www.accf.org/pdf/test-kyoto-oct52005.pdf.

15. Data found at http://www.accf.org/pdf/ACCF_KyotoEconImp.pdf

16. *General equilibrium* models, in contrast, take into account the effects of emissions restrictions on other economic sectors. These show much greater negative economic effects than sectoral models in terms of

job losses and reduced growth. See, Canes, M., *Economic Modeling of Climate Change Policy*, International Council for Capital Formation, October 2002.

17. "Evaluating the role of prices and R&D in pricing carbon dioxide emissions," Congressional Budget Office, September 2006, found at: http://usinfo.state.gov/products/pubs/oecon/chap3.htm.

18. Press Release, "EU15 greenhouse gas emissions increase for second year in a row," European Environment Agency, July 22, 2004 at http://org.eea.europa.eu/documents/newsreleases/GHG2006-en.

19. See, *e.g.*, Transcript, *The News Hour with Jim Lehrer*, August 4, 2006, found at http://www.pbs.org/newshour/bb/weather/july-dec06/climate_08-04.html.

20. The economy is dynamic and therefore differently subject to price hikes at different times; but for a brief discussion of the slowing and inflationary impacts of energy price increases see, *e.g.*, "The U.S. Economy: A Brief History", U.S. Department of State, at http://usinfo.state.gov/products/pubs/oecon/chap3.htm.

21. Ibid., chapter found at http://www.cbo.gov/showdoc.cfm?index=2876&sequence=4.

22. Cooler Heads Newsletter, Nov. 12, 2003. See http://www.global-warming.org/article.php?uid=233.

23. This conclusion, as is typically the case in science, is not universal. For these purposes, we note the findings but do not feel compelled to rely on the possibility, given ethanol's myriad absurdities as a fuel.

24. Schleede, G. 2004, "Facing up to the True Costs and Benefits of Wind Energy," paper presented to the owners and members of Associated Electric Cooperative, Inc., at the 2004 Annual Meeting in St. Louis, Missouri. Available at http://www.globalwarming.org/aecifa.pdf.

25. The firestorm President Bush caused by repeating the description of nuclear as "renewable" in October 2006 testifies to the greens' continuing fear of the atom far outweighing their fear of "global warming."

26. J. Hollander, *The Real Environmental Crisis: Why Poverty, Not Affluence, is the Environment's Number One Enemy*, April 2003, available at https://www.ucpress.edu/books/pages/9208.html.

27. See "Living with Global Warming," September 14, 2005, found at http://www.ncpa.org/pub/st/st278/.

28. See "No Way Back: Why Air Pollution Will Continue to Decline," June 11, 2003, found at: http://www.aei.org/books/filter.all,bookID. 428/book_detail.asp.

29. See Energy Information Administration, Office of Integrated Analysis and Forecasting, U.S. Department of Energy, "Analysis of Strategies for Reducing Multiple Emissions from Power Plants: Sulfur Dioxide, Nitrogen Oxides, and Carbon Dioxide," December 2000, p. xviii, found at: http://www.eia.doe.gov/oiaf/servicerpt/powerplants/pdf/sroiaf(2000)05.pdf.

30. Iain Murray, "What Will We Do When America's Lights Go Out?" *Washington Examiner*, November 13, 2006.

Chapter 12
The Kyoto Protocol

1. For list of signatory and ratifying countries see http://unfccc.int/resource/conv/ratlist.pdf.

2. See, *e.g.,* http://unfccc.int/resource/docs/natc/eunc3.pdf. The EU, which under Kyoto negotiated a "bubble" such that it could pool its increases and "reductions," announced that it met its Rio target. It said that by the benchmark of 2000 it had reduced greenhouse gases by 3.5 percent below 1990 levels. This directly and solely due to the ending of coal subsidies in Great Britain in their push to replace coal with gas, shutting down East German industry, both of which political decisions pre-date and have nothing to do with Kyoto, Rio or other "global warming" pact. Also, note that Europe did not match the U.S.'s decade-long economic expansion. Russia, *e.g.,* met its target by regressing economically.

Regardless, in the build-up to the August 2002 World Summit on Sustainable Development in Johannesburg, Senator James Jeffords (I-Vt.) chaired a joint hearing by the Senate Committees on Foreign Relations on Environment and Public Works to address the issue of U.S. failure to meet its commitments under MEAs. The author testified at this hearing (much of which prepared testimony is presented herein and which may be found at http://www.cei.org/gencon/027,03136.cfm) was asked to testify on the basis that Jeffords' argument was that the U.S. failure to meet Rio's commitment made its Kyoto commitments the necessary and logical next step.

3. S.Res. 98 105th Congress (105-54 July 21, 1997).

4. See, *e.g.*, "GLOBAL WARMING ALERT: GORE BURNS 439,500 LBS OF FUEL TO ATTEND SUMMIT," Drudge Report archives, December 7, 1997 http://www.drudgereportarchives.com/dsp/ specialReports_pc_carden_ detail.htm?reportID=%7B08FAD3B1-9FAF-4037-BF26-573B2826B2C3%7D.

5. See, *e.g.*, "Gore pledges 'flexibility' at climate summit", CNN.com, December 8, 1997, found at: http://www.cnn.com/EARTH/9712/08/climate.gore.wrap/index.html.

6. See the EU Burden Sharing Agreement obligations at http://www.climnet.org/resources/euburden.htm.

7. http://www.euractiv.com/en/sustainability/interview-ceps-researcher-christian-egenhofer-eu-us-climate-change-policies/article-140339.

8. See, e.g., "Geologic constraints on global climate variability," Dr. Lee C. Gerhard, Power Point presentation ound at http://ff.org/centers/csspp/docs/gerhardppt.ppt, text at http://ff.org/centers/csspp/pdf/gerhardnotes.pdf.

9. T.M.L. Wigley, "The Kyoto Protocol:CO2, CH4, and climate implications," *Geophysical Research Letters*, Vol. 25, No. 13., 2285-2284, July 1, 1998.

10. James M. Lindsay, "Global Warming Heats Up: Uncertainties, both scientific and political, lie ahead," *The Brookings Review*, Fall 2001, Vol. 19, No. 4.

11. Annual European Community greenhouse gas inventory 1990–2004 and inventory report 2006, http://reports.eea.europa.eu/technical_report_2006_6/en/EC-GHG-Inventory-2006.pdf.

12. http://www.statistics.gov.uk/cci/nugget.asp?id=901.

13. See, "2004 UK climate change sustainable development indicator and greenhouse gas emissions final figures," January 23, 2006, found at: http://www.defra.gov.uk/news/2006/060123b.htm.

14. http://www.iea.org/textbase/press/pressdetail.asp?PRESS_REL_ID=80.

15. http://reports.eea.eu.int/technical_report_2004_7/en/Analysis_of_GHG_trends_and_projections_in_ Europe.pdf.

16. December 2005 EEA press release announcing their annual numbers, at: http://europa.eu.int/rapid/ pressReleasesAction.do?reference=IP/05/1519&format=HTML&aged=0&language=EN&guiLanguage=en.

17. "Greenhouse gas emission trends and projections in Europe 2005," EEA Report No 8/2005, found at: http://reports.eea.europa.eu/ eea_report_2005_8/en.

18. Source: European Environment Agency, June 2006, available at: http://org.eea.europa.eu/documents/newsreleases/GHG2006-en.

19. http://www.eia.doe.gov/pub/international/iealf/tableh1co2.xls, last updated July 2006.

20. See, *e.g.*, http://europa.eu.int/rapid/pressReleasesAction.do?reference=IP/05/1519&format=HTML &aged=0&language=EN&guiLanguage=en.

21. See, *e.g.*, http://www.eia.doe.gov/pub/international/iealf/tableh1co2.xls.

22. See "Notes and Definitions" on GDP, found at: https://www.cia.gov/cia/publications/factbook/docs/notesanddefs.html.

23. See, *e.g.*, "Katrina Should Be a Lesson to US on Global Warming," *Speigel Online*, August 30, 2005, found at http://www.spiegel.de/international/0,1518,372179,00.html, referring to and translating Tritten's piece in the *Frankfurter Rundschau*.

24. See letter notifying the UN of U.S. withdrawal, or the "unsigning," at http://www.state.gov/r/pa/prs/ps/2002/9968.htm.

25. Reuters, "States Tackle Warming," http://www.ocregister.com/ocr/2005/08/25/sections/news/news/article_647450.php.

26. Katherine Griffiths, "Exxon Shareholders Call for More Details on Kyoto," *The Independent*, May 26, 2005.

27. http://www.johnkerry.com/pressroom/speeches/spc_2003_0123.html.

28. http://www.gwu.edu/~action/2004/issues/kerr020903spenv.html.

29. *FOX News Sunday*, August 2, 2004, transcript at: http://www.foxnews.com/story/0,2933,127599,00.html.

30. http://www.debates.org/pages/trans2004c.html.

31. http://www.johnkerry.com/pressroom/speeches/spc_2003_0123.html.

32. "Remarks of Senator Hillary Rodham Clinton At The National Press Club On Energy Policy," May 23, 2006, found at: http://clinton.senate.gov/news/statements/details.cfm?id=255982.

33. For a discussion of how Europe chased the U.S. away from Kyoto, under the Clinton administration, see C. Horner, "Senator Daschle (D., EU):

Dakotan Angry That U.S. Decides Its Own Positions," *National Review Online*, October 21, 2002.

34. See collection of relevant discussion on this development, "Reflections on the collapse of COP-6", December 2, 2000, Science and Environment Policy Project, found at http://www.sepp.org/Archive/weekwas/2000/dec2.htm.

35. See, e.g., "Who killed Kyoto?" Iain Murray, TechCentralStation.com, May 30, 2003, found at http://www.cei.org/gencon/019,03479.cfm.

36. B. Gwertzman, "Europe Is Betting Against Bush," interview with Guillaume Parmentier, director of the Center on the United States at the French Institute for International Relations, October 14, 2004, found at: http://www.cfr.org/publication/7450/.

INDEX

A

AASC. *See* American Association of State Climatologists
ABC, 107, 160, 173, 177, 178
abortion, 10, 30
Access to Energy, 7
Acerinox, 259
acid rain, xiii, 261
AEI. *See* American Enterprise Institute
Agence France-Press, 155
Agriculture Department, U.S., 235
AI: Artificial Intelligence, 188–89
AIDS, 26, 52
AIT. See An Inconvenient Truth
ALA. *See* American Lung Association
alarmism, climate: big business and, 4, 193–207; causes of, 46; climate models and, 71, 131–33, 136–38; costs of, 245–70; energy, alternative sources of and, 37; environmentalism and, 37–59; EPA and, 42–43; global cooling and, 37, 169, 171; global warming and, 39, 51–53, 55–59, 78; gloom, desire for and, 49–51; Hollywood and, 187–91; hurricanes and, 175; media and, 41–42, 47–49, 169–91, 182–86; pollution and, 39, 40, 44–45; positive trends and, 37–39
Alaska, 132, 170–71, 182
Alaska Climate Research Center, 132
Alaska Monthly Summary, 182
Albania, 11
Alexander, Will, 225
Allen, Myles, 32
Alter, Ethan, 211
American Association of State Climatologists (AASC), 85

American Enterprise Institute (AEI), 17, 41, 94
American Gas Association, 198
American Geophysical Union, 110, 156
American Lung Association (ALA), 42
American Meteorological Society, 157, 159
The American President, 190
American Wind Energy Association, 198
Amery, Carl, 283
Ammann, Caspar, 88
Amoco. *See* British Petroleum
Animal Farm (Orwell), 24
Antarctica, 127, 149–50, 149fig, 231–32
Anthes, Richard, 151
anti-Americanism, 7, 271, 295
Anti-Americanism (Revel), 31
AP. *See* Associated Press
Archer Daniels Midland, 201, 254, 267
Arctic: dragonflies and, 171–72; *An Inconvenient Truth* and, 230–31; media and, 174–76; solar activity and, 146fig, 225fig; temperatures in, 146fig; warming in, 26, 66–67, 143–46, 218
Arctic Climate Impact Assessment (ACIA), 144–45
Århus, University of, Denmark, 39
Arrhenius, Svante, 96
Associated Press (AP), 38, 101, 130, 142, 161, 294
Attenborough, Sir David, 9–10
Auken, Sven, 286
avian flu, 26, 169

B

Baden, John A., 213
Bailey, Ron, 26, 99

Ball, Tim, 230
Balling, Robert C., 223, 236
Bambi, 181, 187
Bangladesh, 33
Barton, Joe, 128
BBC. *See* British Broadcasting Corporation
Beam, Alex, 178–79
Bellamy, David, 267
Bennett, Catherine, 294
Bening, Annette, 190
big business: alarmism, climate and, 193–207; British Petroleum, 199–201; Cinergy, 203–5; DuPont, 202–3; Enron, 194–99; environmentalism and, 3, 4, 4–6, 193–207; green alarmism and, 4; Kyoto Protocol and, 64, 193, 194–96; media and, 205–6
The Big Ripoff (Carney), 200
Bing, Steve, 206–7
Bingaman, Jeff, 249–50, 251
biomass, 59
Blair, Tony, 51
Blakemore, Bill, 173, 178
Blyth, William, 261
Boehlert, Sherwood, 128
Bolles, Edmund Blair, 174
Boston Globe, 130, 178
Botswana, 11
Boulding, Kenneth, 31
BP. *See* British Petroleum
Bray, Dennis, 83
Brazil, 53, 265
British Antarctic Survey, 10
British Broadcasting Corporation (BBC), 10, 26, 29, 142, 184–85, 235
British Petroleum (BP), xvi, 93, 195, 199–201, 276
Brokaw, Tom, 180
Brookings Institution, 284
Brown, Gordon, 288
Brown, Lester, 49–51, 59
Browne, John, xvi, 196, 199, 200, 276
Browner, Carol, 37
Bruno, Giordano, 225
Brynner, Rock, 188
Bryson, Reid, 55
Burke, Edmund, 9
Burleigh, Peter, 277
Bush, George H. W., 273
Bush, George W., 35, 38, 146; Clear Skies Initiative of, 40; consensus and, 101; energy policy of, 59, 196; environmental policy of, 40–41, 196; Kyoto Protocol and, 64, 65, 275, 277, 292–95, 295–98; science and, 81, 108
Business Council for Sustainable Energy, 198
Business Environmental Leadership Council, 194
Business and Media Institute (BMI), 182–83
Byrd, Robert, 275

C

Calder, Nigel, 55
Calgary Herald, 35, 177
Calpine, 197
Canada, 177, 257
capitalism, environmentalism and, 3, 5–6, 7–9, 13, 15–17, 19, 20
carbon dioxide emissions: arctic temperatures and, 146fig; climate-change litigation and, 31–32; energy use and, 245; in Europe, 283–89; global warming and, 45, 62–63, 65, 66, 68–70, 73, 79, 82, 115, 220–22; IPCC and, 114; Kyoto Protocol and, xvi, 46; in Northern Hemisphere, 114; in Southern Hemisphere, 114; temperature and, 114–15, 216–17
Carleton University, 221
Carney, Timothy P., 200
Carson, Rachel, 9
Carter, Bob, 91, 186, 219
Carter, Jimmy, 20
Cato Institute, 11
CBO. *See* Congressional Budget Office
CBS, xvi, 177
Centers for Disease Control (CDC), 238
Centre for Resource Management Studies, 105
Centre for Sun-Climate Research, 79
Charles River Associates, 258
Chautauqua Institution, 179
Cheney, Dick, 190
China, 8, 53, 117, 257, 265, 281, 299
The China Syndrome, 188
Chirac, Jacques, xiii, 6, 193, 300, 301
chlorofluorocarbons (CFCs), 99
Christian Science Monitor, 210
Christy, John, 71
Churchill, Ward, 94
Churchill, Winston, 210, 240–41, 247

Chylek, Petr, 38, 228–30
Cinergy, 93, 203–5
Citizens for a Sound Economy (CSE), 96
Clean Air Act, 45, 261, 267
Clean Air Task Force, 197
Clean Power Group, 197–98
Clear Skies Initiative, 40
climate: feedback and, 233–34; global history of, 118–19; global warming and, 65; hurricanes and, 150–52; litigation and, 30–33; man's contribution to, 82, 95–96, 226; people and, 11; poverty and, 64; sea levels and, 63–64; solar activity and, 225–28; stability of, 63, 66; weather vs., 64, 173
Climate and Economy Insurance Act, 251
ClimateAudit.org, 82
climate models: alarmism, climate and, 131–33, 136–38; Antarctic cooling and, 127; benefits of, 133; global warming and, 131–38; IPCC and, 136; media and, 137; problems of, 133; sea levels and, 131; temperature and, 131, 134–35; testing, 133–34; unreliability of, 113
Climate Science Watch, 108
Climate Stewardship Act, 250–51
Clinton, Bill, xvi, 190, 255, 297; Enron and, 196–97; environmentalism and terrorism and, 4; environmental policy of, 40–41; global warming alarmism and, 51–52; Hockey Stick scandal and, 122; Kyoto Protocol and, xvi, 64, 184, 271, 277, 278, 287, 292–93, 298
Clinton, Hillary, 275, 297, 298
CNN, 109, 130, 215
CNSNews.com, 32, 187
CO$_2$ emissions. See carbon dioxide emissions
CO2Science.org, 150
Cold War, 7
Cole, Lamont, 15
Colligan, Douglas, 98
Columbia Review of Journalism, 35
Columbia University, 188
Committee for a Constructive Tomorrow (CFACT), 33
Commons Committee on Environment and Sustainable Development, 221
communism, 3, 7
Competitive Enterprise Institute, 34, 35, 205, 250

Congressional Budget Office (CBO), 255, 260
Congressional Record, 249
consensus: Bush, George W. and, 101; character assassination and, 81, 87–89, 95–99, 99–106; discredited reports and, 81, 82–87, 89–94; fake experts and, 81, 106–10; global cooling and, 77, 81; global warming and, xv, 65, 67, 81–110, 218; Gore, Al and, 108–10; An Inconvenient Truth and, 89; IPCC and, 94, 103, 110; media and, 81; NAS and, 94; politics and, 81; scientific method and, xv
conservationism, 3, 9, 13, 23
"Convention on the Protection of the Environment through Criminal Law," 30
Cooney, Phil, 100–103
Copenhagen Post, 79
Copernicus, 88
Corcoran, Terence, 85–86, 108, 177
Cosmos journal, 97
Costner, Kevin, 188
Council on Sustainable Development, 197
Countdown to Doomsday (MSNBC), 8
Crichton, Michael, 13, 18
Curry, Judith, 151

D

Danish Ministry of Science, Technology and Innovation, 105
Danish National Space Center, 79
Daschle, Tom, 293, 298, 300–301
David, Laurie, 46
The Da Vinci Code (Brown), 122
Davis, Angela, 7
The Day After Tomorrow, 190, 215, 232
DDT, 17, 53, 236
Deming, David, 119
Denmark, 39, 285–86
Desert Storm, Operation, xvi
Discover magazine, 6, 40
Discovery Channel, 180
diseases, vector-borne, 237–39
Doran, Peter, 127
dragonflies, 171–72
Driessen, Paul, 19
drought, 37, 141, 161–62
Duke Energy, 93, 203, 206. See also Cinergy
DuPont, 202–3, 206, 254
Dyson, Freeman, 113

E

Earth in the Balance (Gore), 95, 97, 99, 100, 187, 210–11, 221

Earth First!, 16

Earth Policy Institute, 50

East Anglia, University of, 219

Easterbrook, Gregg, 85, 97; alarmism, climate and, 185–86; Arctic melting and, 175; character assassination of, 104–05; global warming and, 48–49; global warming solutions and, 260–62, 268

Ebell, Myron, 205, 209n

Ebert, Roger, 211–12

Eco-Imperialism: Green Power, Black Death (Driessen), 19

ecology, 17, 18

"Ecomagination" campaign, 206

Economist, 181–82

economy: energy and, xiv, 78, 245, 263–64; environmentalism and, 22; GHGs and, 139; global cooling and, 28; global warming solutions and, 65; government control and, 3–4, 26, 28; IPCC and, 138–39

Ecotage, 16

ecotourism, 142–43

EDF. *See* Environmental Defense Fund

Egenhofer, Christian, 278

Ehrlich, Paul, 4, 10, 27, 39, 49, 51, 212, 234; alarmism, climate and, 45; IPAT formula and, 11; pollution and, 45; population and, 41; U.S. economic growth and, 11

Eisenhower, Dwight D., 303

El Niño, 113, 116

El Paso, 197

Elsner, James, 151

Emanuel, Kerry, 148, 151, 152

Emergency Disasters Database, 164

The End of Affluence (Ehrlich), 51

energy: alternative sources of, 37, 55–56, 58–59, 199, 247, 266–67; carbon dioxide emissions and, 245; cheap, abundant, 4, 5, 25; cost of, 26; demand for, 80, 246–47; economy and, xiv, 78, 245; Kyoto Protocol and, 76, 78–80; rationing of, xv, 6, 29, 193; renewable, 55, 239–41, 266–67; technology and, 79; terrorism and, 4

Energy Department, U.S., 99–100, 197, 252, 255–58

Energy Information Administration, 252, 268, 290

Enron, 93, 193, 194–99; Bush, George W. and, 196; Clinton, Bill and, 196–97; Gore, Al and, 197; Kyoto Protocol and, 194–96, 199

Environmental Defense, 181, 197

Environmental Defense Fund (EDF), 17, 104–5

Environmental Gore (ed. Baden), 213

Environmental News Service, 67

Environmental Protection Agency (EPA), 37, 42–43, 47, 68, 100, 132, 197

environmentalism: agenda of, 3–24; alarmism, climate and, 37–59; as anti-American, 7–9, 23; as anti-capitalist, 3, 5–6, 7–9, 13, 15–17, 19, 20, 23; big business and, 3, 4, 4–6, 193–207; capitalism and, 3; dissent and, 4–5, 34–35, 81; government control and, xiii, 3–4, 12–13, 25–30; mental health and, 54; naturalism and, 22; people, antipathy toward and, 9–11, 23–24; population and, 10, 22; positive trends and, 37–39; for profit, 193–207; as religion, 18; terrorism and, 3, 4, 21; UN and, 6

EPA. *See* Environmental Protection Agency

Eskimos, Inuit, 16–17, 172

Esper, Jan, 127

EU. *See* European Union

Europe: 2003 heat wave in, 219–20; anti-Americanism in, 271; carbon dioxide emissions of, 283–89; gas prices in, 20–21; GHGs and, 65; global warming solutions and, 258–60; Hurricane Katrina and, 76; Kyoto Protocol and, 65, 258–60, 271–72, 283–89

European Emissions Trading Scheme, 261

European Environment Agency, 271, 284–85

European Parliament, 21, 49, 76, 165

European Union (EU), 6, 12, 63, 149; alarmism, climate and, 50; energy, alternative sources of and, 59; Kyoto Protocol and, 261, 281, 286–89

Exxon-Mobil, 201

Exxon Valdez, 188

F

Facts, Not Fear (Sanera and Shaw), 6

Fagan, Brian, 237

Fallow, Brian, 248

The Fate of the Earth (Schell), 56

Federal Bureau of Investigation (FBI), 3
Federation of American Scientists Public Issue Report (1978), 4
Financial Post (Canada), 35, 108
Financial Times, 200, 201, 205, 206
flooding, 141, 162–65
Ford, Henry, 267
Foreign Affairs, 298
Fortune magazine, 170, 189
fossil fuels, 22, 59; global warming and, 49; temperature and, 116–17; U.S. use of, 116
Fourth Assessment Report (IPCC), 121
FOX News Channel, 180
FOX News Sunday, 296
Frankenstein, 188
Freddoso, David, 252–53
Freeh, Louis B., 14
Freshfields (law firm), 30–31
Friedan, Betty, 22
Friedman, Milton, 25, 26
Friends of the Earth, 21, 26, 57
Fundacion Argentina de Ecologia Cientifica, 223

G

Gabon, 11
Galileo Galilei, 88, 225
Gallup, 38
Galst, Liz, 54
gas prices, 20–21, 257
Gas Turbine Association, 198
Gazprom (Russia), 195
general circulation models (GCMs). *See* climate models
General Electric (GE), 93, 194, 201, 206
George Mason University, 132
Germany, 152, 165, 284, 286, 292
GHGs. *See* greenhouse gases
Giegengack, Robert, 90
GKSS Forschungszentrum, 83
glacial melting, 63, 65, 66–67, 76–77, 157; climate models and, 131; *An Inconvenient Truth* and, 223–28; sea levels and, 147–49
global cooling, 15; alarmism, climate and, 37, 169, 171; consensus and, 77, 81; economy and, 28; glacial melting and, 222; global warming myths and, 62; government control and, 28; media and, 81, 169, 171; pollution and, 55; U.S. and, 116

global governance, xiii, 6, 271, 275, 299–301
Global Insight, 259
Global Salvationist movement, 7, 15–17, 212
global warming: alarmism, climate and, 39, 78; benefits of, 65, 66, 75; carbon dioxide emissions and, 45, 62–63, 65, 66, 68–70, 73, 79, 82, 115; as catastrophic, 65, 72, 87; causes of, 32, 67–70, 90; climate change and, 65; climate-change litigation and, 30–33; climate models and, 71, 131–38; consensus and, xv, 65, 67, 81–110, 218; effects of, 67, 73–75, 212–14; energy demand and, 246–47; environmentalism and, xiii; extent of, 66, 67, 70–73; GHGs and, 21, 48, 54–55, 62–63, 66, 68–70, 72–73, 82, 214–17; glacial melting and, 63, 65, 66–67, 76–77, 157, 223–28; government control and, xiv–xv, 81; Hockey Stick scandal and, 70, 88, 119–31; Hollywood and, 187–91; hurricanes and, 76, 141, 148, 155, 158–61, 162, 215; *An Inconvenient Truth* and, 209–41; as manmade, 65, 66, 68–70, 87; media and, 77–78, 169–91; myths of, 62–64; terrorism and, 52; UN and, xiii, 63; U.S. and, 64. *See also* global warming solutions; Kyoto Protocol
global warming solutions: actual impact of, 247, 250–52, 264–65; alarmism, climate and, 54–55, 55–59; carbon dioxide emissions and, 245–46; carbon rationing and, 247–50; cost/benefit analysis and, 250–52; costs of, 78–80, 245–46, 267–69; economic effects of, 255–58; energy demand and, 246–47; Europe and, 258–60; government control and, xiv–xv; poverty and, 263–64; technology and, 265–67. *See also* Kyoto Protocol
globalization, 20
Globe and Mail (Canada), 177
Goklany, Indur, 268
Goldenberg, Stanley, 156
Good Morning America, 179
Gore, Al, xvi, 34, 37, 46, 52, 53, 81, 93, 153, 179; alarmism, climate and, 162–65, 174, 209–10; character assassination and, 95–99, 99–106; climate models and, 137–38; consensus and, 89–90, 93–94, 95–99, 99–106, 108–10; drought and, 161, 162;

Gore, Al (*con't*), *Earth in the Balance* of, 187, 210–11, 221; Enron and, 193, 197; flooding and, 162–65; global cooling and, 219; global warming and, xiii, xiv, 63, 71, 174, 193, 210, 212–14; Hockey Stick scandal and, 122, 123, 221–22; *An Inconvenient Truth* and, 162–65, 190–91, 209–41; Kilimanjaro, Mount and, 209, 213, 218; Kyoto Protocol and, xvi, 64, 194, 209, 252, 271, 278, 293, 301; Oreskes, Naomi and, 89–90, 93–94; polar bears and, 141, 142; Revelle, Roger and, 95–99; Rio treaty and, 272–74; sea levels and, 147, 149, 150, 209; storm activity and, 64. *See also An Inconvenient Truth*
Gore, Tipper, 174
government control: economy and, 3–4, 26, 28; environmentalism and, xiii, 3–4, 12–13, 25–30; global cooling and, 28; global warming and, xiv–xv, 28, 81; individual activity and, 3–4; pollution and, xiii; population and, 27
Gray, William, 108, 108–10, 155, 162
Great Alaskan Warming, 170–71
Green Party, 7, 121
greenhouse effect, 79, 187, 217, 233, 280
greenhouse gases (GHGs): Arctic temperatures and, 145–46; economic growth and, 139; global warming and, 21, 48, 54–55, 62–63, 66, 68–70, 72–73, 82, 214–17; hurricanes and, 141, 150, 159; hydroelectric power and, 55–56; industrial activity and, 82; Kyoto Protocol and, 271, 273–74; man's contribution to, 68–70, 69fig, 72, 137; reduction of, 49; sea levels and, 209; temperature and, 117; trees, use of and, 56–58
The Greening of Planet Earth: The Effects of Carbon Dioxide on the Biosphere, 237
Greenland, 66, 77, 120, 147, 149–50, 216, 228–29
Greenpeace, 7, 40, 43, 181–82, 223
greens. *See* environmentalism
Grinnell Glacier, 222fig, 233
Griscom, Amanda, 104
Grist magazine, 34–35, 104, 186, 214
The Guardian (UK), 34, 48, 294
Gulf Stream shutdown, 232–33

H

Hadley Centre, UK, 134

Hagel, Chuck, 275
Hansen, James, 52, 84, 95, 109, 181; alarmism, climate and, 136–37; global warming and, 32–33; *An Inconvenient Truth* and, 210; media and, 180; sea levels and, 147, 232
Happer, William, 99–100
Hard Green: Saving the Environment from the Environmentalists (Huber), 9, 75
Hardin, Garrett, 27
Harper's, 22
Harris, Tom, 236
Harvard University, 114, 201
Hayek, Friedrich, 25
Hayward, Steven, 17, 23, 38, 56, 139
Hazelwood, Joseph, 188
HBO, 180
The Heartland Institute, 96
Heinz Center for Science, Economics and the Environment, 194, 198
Heinz, Teresa, 181, 198
heliocentrism, 225
Henderson, David, 17
Hiss, Alger, 5
Hitler, Adolf, xvi
"Hockey Stick" graph, 70, 88
Hockey Stick scandal, 119–31; Clinton, Bill and, 122; Gore, Al and, 122, 123, 221–22; IPCC and, 125–26; Little Ice Age and, 120, 124, 126–27, 129, 130; Mann, Michael and, 120, 122–24, 125–31; media and, 129–30, 177; NAS and, 122, 128–31; temperature readings and, 122–23; Thousand-Year Spike and, 129; UN and, 128
Hoffert, Martin I., 246–47, 257
Holland, Greg, 151
Hollander, Jack, 247, 268
Hollywood, 3, 19, 149; alarmism, climate and, 187–91
Hopper, Dennis, 188
Huber, Peter W., 9, 75
Hughes, Warwick, 133
Hulme, Mike, 46
Humlum, Ole, 182
Hurricane Andrew, 158
Hurricane Katrina, 12, 50, 76, 141, 148, 156, 175, 236, 237, 292
hurricanes: in 2006, 141, 150, 151; alarmism, climate and, 175; climate change and, 150–52; as cyclical, 181; GHGs and, 141, 150, 159; global warm-

ing and, 76, 141, 148, 155, 158–61, 162, 215; IPCC and, 154–55, 159; man's contribution to, 154–55; population and, 151; trends of, 156–59
Hussein, Saddam, xvi, 26
hydroelectric power, 37, 55–56, 58–59

I

ICC. *See* International Criminal Court
An Inconvenient Truth (AIT), 91, 95; alarmism, climate and, 162–65; Antarctic melting and, 231–32; Arctic melting and, 230–31; consensus and, 89; diseases, vector-borne and, 237–39; *Earth in the Balance* (Gore) and, 210–11; extinctions and, 234–35; global warming, effects of and, 212–14; Gulf Stream shutdown and, 232–33; melting ice sheets and, 228–41; misrepresentations in, 218–28; reviews of, 211–12; sea levels and, 209, 232; sins of omission in, 214–17; solutions and, 239–41; species migration and, 236; storm activity and, 236–37. *See also* Gore, Al
The Independent, 34, 179, 294
Independent Television Network (ITN), 101–3
The Index of Leading Environmental Indicators: The Nature and Sources of Ecological Progress in the U.S. and the World (Hayward), 17, 38
India, 53, 79, 117, 202, 257, 265, 299
Industrial Revolution, 70, 79, 119, 158
Inhofe, James, 84, 130, 179
insects, 141, 171–72
Inside Politics, 109
Institut Pasteur, 238
Intergovernmental Panel on Climate Change (IPCC), 63, 83, 84; Arctic melting and, 218, 230–31; carbon dioxide emissions and, 114; climate models and, 136; consensus and, 83, 84, 94, 103, 110; economic growth and, 138–39; Fourth Assessment Report of, 121; Hockey Stick scandal and, 125–26; hurricanes and, 154–55, 159; sea levels and, 148–49; storm activity and, 64, 236–37; Summary for Policymakers of, 84, 120, 148, 236; temperature history and, 119; Third Assessment Report of, 103, 120, 148, 159
Interior Department, U.S., 268
International Arctic Research Center, 163

International Criminal Court (ICC), 30
International Energy Agency, 80, 261, 281, 285
International Herald Tribune, 261
International Journal of Climatology, 223, 227
International Policy Network, 83
International Wildlife, 55
Interstate Natural Gas Association, 198
Inuit Eskimos, 16–17, 172
IPCC. *See* Intergovernmental Panel on Climate Change
ITN. *See* Independent Television Network

J

James Cook University, Australia, 91
Jenkins, Simon, 26
John Locke Foundation, 43
Jones, Phil, 133
Journal of Environmental Education, 237
Journal of Geophysical Research Letters, 88, 223, 228

K

Kant, Immanuel, 18
Karlén, Wibjörn, 231
Kaser, G., 227
Kelly, Petra, 7
Kenya, 63
Kerry, John, 130, 181, 206, 252; Kyoto Protocol and, 295–98
Khosla, Vinod, 206
Kilimanjaro, 63, 209, 213, 218, 222–23, 227
King, Sir David, 51–52, 109
Kirk, Russell, 9
Kissinger, Henry, 97
Klotzbach, Phil, 151, 159–60
Knutson, Tom, 151
Kolbert, Elizabeth, 186
Koppel, Ted, 102
Krenicki, John, 206
Kristof, Nicholas, 57
Kyoto Protocol, 21, 45; actual impact of, 264–65, 272, 279–83; big business and, 64, 193, 194–96; Bush, George W. and, 64, 65, 275, 277, 292–95, 295–98; cap-and-trade quota system and, 255, 278–79; carbon dioxide emissions and, xvi, 46; climate-change litigation and, 31–32; Clinton, Bill and, xvi, 64, 184, 271, 277, 278, 287,

Kyoto Protocol (*con't*), 292–93, 298; costs of, 267–69, 271; economic effects of, 255–58, 261; effectiveness of, 82; effects of, 39; energy rationing and, 78–80; energy use and, 76; Enron and, 194–96, 199; EU and, 281, 286–89; in Europe, 65, 258–60, 271–72, 283–89; flaws of, 278–79; GHGs and, 54–55; global governance and, xiii, 6, 271, 275, 299–301; global warming and, 252–55; goal of, 273–74; Gore, Al and, xvi, 64, 194, 209, 252, 271, 278, 293, 301; greenhouse gases (GHGs) and, 271, 273–74; implementation of, 67; inadequacy of, 52–53; purpose of, 64; ratification of, 271; Senate, U.S. and, 276–77; United States and, xvi, 64, 292–95, 301. *See also* global warming

L

Landsea, Christopher, 151, 152, 154, 156
Laos, 11
Lauer, Matt, 8–9
Lautenbacher, Conrad, 160
lawn care industry, xv
Lay, Ken, xvi, 194, 196, 197, 198, 276
League of Conservation Voters, 10
Lenin, Vladimir, 171
Lewis, Marlo, 209n, 250–51
Lieberman, Joe, 250, 277, 297
Lincoln, Abraham, 210
Lindzen, Richard, 59, 77, 83–85, 86, 93, 154, 235
litigation, climate-change, 30–33
Little Ice Age, 62, 66, 70, 72, 77, 111, 119, 145; Hockey Stick scandal and, 120, 124, 126–27, 129, 130
The Little Ice Age: How Climate Made History (Fagan), 237
Lomborg, Bjorn, 38–49, 52, 105–6, 109, 268, 299
London Mail on Sunday, 181
Los Alamos National Laboratories, 228
Lovelock, James, 23–24
Lovins, Amory, 5
Ludlum, Robert, 190
Lynas, Mark, 35
Lyons, Rob, 133

M

Madonna, 16
Mahoney, James, 204

malaria, 53, 238–39
Malawi, 16
Maldives, 149
Malthus, Thomas, 24, 49, 212, 213
Mann, Michael, 120, 122–24, 125–31
Marine Geophysical Laboratory, Australia, 91
marriage, 29, 30
Marx, Karl, 18
Massachusetts Institute of Technology (MIT), 148
Massachusetts, University of, 224
Max Planck Institute, 58
May, Lord Robert, 104
Mayfield, Max, 107, 151, 175
McCain, John, 249, 250, 259, 277, 288, 297
McCarthyism, 128
McGinty, Katie, 99
McIntyre, Steven, 124–27
McKenna, Malcolm, 174
McKitrick, Ross, 124–27, 219
McNamara, Robert, 27
McShane, Owen, 105, 179
media, 3; alarmism, climate and, 41–42, 47–49, 81, 169, 169–91; Arctic melting and, 174–76; balanced reporting and, 169, 176–82; big business and, 205–6; climate models and, 137; communism and, 7; consensus and, 81; global cooling and, 81, 169, 171, 182–86; global warming and, 77–78, 169–91; Hockey Stick scandal and, 129–30, 177; hurricanes and global warming and, 159–61; inconsistency of, 182–87; people as pollution and, 11; pollution, air and, 169–70; positive trends and, 38
Media Research Center, 173
Medieval Climate Optimum, 70, 124
Medieval Warm Period, 66, 111, 119, 120, 126–27, 129
Meltdown: The Predictable Distortion of Global Warming by Politicians, Science and the Media (Michaels), 143
Mendelson, Robert, 75
Mensch & Energie, 283
mental health, environmentalism and, 54
Merali, Zeeya, 58
methane, 68
Mexico, 257, 265, 299
Meyer, Hans, 227
Michaels, Patrick, 135–36, 143
Middle East, 20, 295

MIT. *See* Massachusetts Institute of Technology

Mitos y Fraudes, 223

Mitra, Barun, 187

Moldova, 11

Moler, Betsy, 197

A Moment on Earth (Easterbrook), 104–5

Monbiot, George, 28, 33

Moore, Patrick, 89

Morgan Stanley, 206

Mörner, Nils-Axel, 148, 149

Morris, Julian, 83

Motl, Lubos, 114, 130–31

MoveOn.org, 193

Moyers, Bill, 34, 102

Mozart, Wolfgang Amadeus, 18

MSNBC, 8, 175

Murray, Iain, 209n, 269–70

Myers, Norman, 234

N

NAAQS. *See* National Ambient Air Quality Standard

NACC. *See* National Assessment on Climate Change (2002)

Namibia, 11

NAS. *See* National Academy of Sciences

NASA, 147, 219

National Academy of Sciences (NAS), 62, 83, 84; Arctic melting and, 218; consensus and, 83, 84, 94; Hockey Stick scandal and, 122, 128–31

National Ambient Air Quality Standard (NAAQS), 45

National Assessment on Climate Change (2002) (NACC), 134

National Center for Policy Analysis, 268

National Geographic, 142, 146

National Hurricane Center, U.S., 107, 108, 156

National Oceanic and Atmospheric Administration (NOAA), 71, 107, 116, 152; global warming and, 173; hurricanes and, 156–58, 160

National Post, 177

NATO. *See* North Atlantic Treaty Organization

Natural Resources Defense Council, 40, 197

naturalism, environmentalism and, 22

Nature magazine, 58, 120, 125, 131, 136

Nelson, Robert, 18

New Orleans, La., 26, 151

New Republic, 17, 97

New Scientist magazine, 58, 163, 225

New York magazine, 161

New York Review of Books, 84, 210

New York Times, 6, 27, 42, 48, 57, 95, 97, 102, 132, 136, 142, 146, 170–71, 174–75, 180, 185, 230, 294

New York University, 161

New Yorker magazine, 186

New Zealand, 257

New Zealand Climate Science Coalition, 105

New Zealand Herald, 248

Newcastle University, 157

Newsweek magazine, 77, 85, 169

Newswire, 89

Newton, Isaac, 18

Nightline, 102

1984 (Orwell), 78

NiSource, 197

NOAA. *See* National Oceanic and Atmospheric Administration

Noah, Timothy, 196

North Atlantic Treaty Organization (NATO), 7

North Pole, 143, 174–76

Northern Hemisphere, 178; carbon dioxide concentrations in, 114; temperatures in, 138fig

Novak, Robert, 196

NOW, 102

nuclear freeze, 56, 190

nuclear power, 37, 55, 58

O

OECD. *See* Organization for Economic Cooperation and Development

Oklahoma, University of, 119

O'Neill, Paul, 198

Operation Desert Storm, xvi

Oppenheimer, Michael, 181

Oregon State University, 144

Oreskes, Naomi, 90–94

Organization for Economic Cooperation and Development (OECD), 17

O'Rourke, P. J., 8

Orrock, John, 23–24

Orwell, George, 24, 78

Oslo University, 182

Oxford Energy Associates, 261

Oxford, University of, 32

ozone hole, xiii
ozone violations, 42fig

P

Pacific Decadal Oscillation, 170–71
Pacific Phytometric Consultants, 236
Palmisano, John, 194, 197
Parmentier, Guillaume, 291
Partisan Review, 18
Pascal, Blaise, 18
Patagonian glaciers, 26
Patterson, Tim, 221
Patzek, Tad, 267
PBS. *See* Public Broadcasting Service
Peiser, Benny, 92
Pelley, Scott, 35, 177
Peña, Federico, 197
Pennsylvania, University of, 90
pesticide industry, xv
Pew Center on Global Climate Change, 194, 202
PG&E National Energy Group, 197
Physics Today, 100
Pielke, Roger, Jr., 159, 163
Piltz, Rick, 108–9
Pimentel, David, 267
Planned Parenthood, 27
Plenty magazine, 54
polar amplification, 143–46, 220
polar bears, 141–43
Policy Statement on Climate Variability and Change, 85
Polissar, P. J., 224–25
Politicizing Science (ed. Gough), 98, 100
pollution, air, 7, 55; alarmism, climate and, 39, 40, 44–45, 47–48; global cooling and, 55; global warming and, 261; government control and, xiii; industrialization and, 55; media and, 169–70; population and, 55; positive trends and, 37; technology and, xiv
Popper, Sir Karl, 124
population: control of, 6; environmentalism and, 10, 22; global warming solutions and, 65; government control and, 27; hurricanes and, 151; pollution and, 55; starvation and, 15; sterilization and, 27; UN and, 6
The Population Bomb (Ehrlich), 11, 41, 49
The Population Explosion (Ehrlich), 11, 49
Potter, Ned, 107

Powell, Colin, 296
Premiere magazine, 211
Pretoria, University of, 225
Princeton University, 100
Progress and Privilege (Tucker), 31
Proposition 87, 206
Public Broadcasting Service (PBS), 34, 102
Purtscheller, Ludwig, 227
Putin, Vladimir, 74
Pythagoras, 88

R

Rapley, Chris, 9–10
The Real Environmental Crisis (Hollander), 247
RealClimate.org, 82, 131, 177
Reason magazine, 26, 99
Reiter, Paul, 238–39
religion, environmentalism as, 18
Resources for the Future, 255
Reuters, 10, 30, 32, 294
Revel, Jean-Francois, 31
Revelle, Roger, 95–99, 100
Revkin, Andrew, 170–71, 183
Ridgeway, James, 17
Rio treaty, 272–74
Rising Tide (Lynas), 35
Ritson, David, 88
The Road to Serfdom (Hayek), 25
Roberts, David, 34
Robertson, Pat, 154
Roth, Sol, 172
Rousseau, Jean-Jacques, 23, 24
Royal Society (RS), 34, 89, 110, 193, 225
Rubin, James P., 298
Russian Academy of Sciences, 74

S

Sallet, Jonathan, 240
Salon.com, 196
Salten, Felix, 187
San Francisco Chronicle, 146
Sanders, Jon, 43
SARS, 26
Scagel, Rob, 236
Schell, Jonathan, 56
Schneider, Stephen, 40, 109, 179
Schröder, Gerhard, 150, 152
Schroeder, Mark, 197
Schwartz, Joel, 41–42, 44, 169, 268–69

science: alarmism, climate and, 85; Bush, George W. and, 81, 108; censorship of, 81; consensus and, 81; energy rationing and, 80; observation and, 88

Science Digest, 98

Science magazine, 27, 75, 78, 83, 93, 156, 160, 221, 246, 247, 257, 265

scientific method, xv, 81, 88, 213

sea levels: climate change and, 63–64; climate models and, 131; GHGs and, 209; glacial melting and, 147–50; global warming and, 65; Gore, Al and, 149, 150, 209; ice cap melting and, 147–49; *An Inconvenient Truth* and, 232; IPCC and, 148–49

Senate Resolution 98, 275–76

September 11, 297

Shakespeare, William, 18

Shifflett, Dave, 174

Shiva, Vandana, 187

Shockley, William, 27

Shuttleworth, John, 21

Sierra Club, 128, 197–98

Silent Spring (Carson), 9

Simon, Julian, 23, 39, 103–4

Singer, S. Fred, 97, 98–99

60 Minutes, 35, 177

Smith, R. J., 50

solar activity: arctic temperatures and, 146fig; 225fig; climate and, 225–28; as cyclical, 181; global warming and, 65

Solar Energy Industry Association, 198

Sontag, Susan, 18

Soon, Willie, 146

South Dakota, 138fig

South Korea, 53, 265

South Pole, 14

Southern Hemisphere: carbon dioxide concentrations in, 114; global warming and, 111, 114–15; temperatures, 114fig

Soviet Union, 7, 70, 74, 116, 245; collapse of, 17, 113; environmental record of, 8

Soylent Green, 172, 187

Spain, 258–59

Spielberg, Steven, 188

Spitzer, Eliot, 33

Stanford University, 185

Stanway, Paul, 218

State of Air report (ALA), 42

State Department, U.S., 197

State of Fear (Crichton), 13

Stehr, Nico, 153

Stephanopoulos, George, 235

Stewart, Christine, 35

Steyn, Mark, 12

"Still Waiting for Greenhouse," 82

St. Louis Journalism Review, 294

Stockholm University, 148, 231

storm activity, 64, 74, 76, 154, 236–37

Stowell, John, 204–5

Strong, Maurice, 6

"Stumbling into War" (Rubin), 298

Summary for Policymakers (IPCC), 84, 120, 148, 236

Sunday Times (UK), 26

Sununu, John, 273

Supreme Court, U.S., 45–46

Svensmark, Henrik, 79

T

Taylor, George, 144

Taylor, Jerry, 11

Taylor, Mitchell, 142

technology: energy use and, 79; global warming solutions and, 265–67; pollution and, xiv; taxation and, 21

The Telegraph (UK), 186

temperature: in Antarctica, 149fig; in Arctic, 143–46; carbon dioxide emissions and, 114–15, 216–17; climate models and, 131, 134–35; fossil fuel consumption and, 116–17; GHGs and, 117; global, 111, 112fig, 118–19, 219fig; history of, 66, 70–72, 122fig; Hockey Stick scandal and, 70, 119–31; measurement of, 70–72, 112–13; in U.S., 115–17

terrorism, 3, 4, 21, 52

Thatcher, Margaret, 278, 284

Third Assessment Report (IPCC), 103, 120, 148, 159

Third World, 18, 245, 260

Three Mile Island, 188

Tierney, John, 27

Time magazine, 78, 141, 161–62, 181, 183–84, 187, 198, 224

Tindale, Stephen, 48

Titanic, 188

tobacco industry, 67

Toronto Star, 177

Toxic Terror: The Truth behind the Cancer Scares (Whelan), 9, 15, 17, 21

Trigen Energy, 197

Trittin, Jürgen, 292

Tucker, William, 31
Twain, Mark, 58
Tyndall Centre for Climate Change Research, 46

U

The Ultimate Resource 2 (Simon), 103
UN Conference on Environment and Development (UNCED), 273
UN Framework Convention on Climate Change (UNFCCC), 273, 275, 285, 286
UNCED. *See* UN Conference on Environment and Development
UNFCCC. *See* UN Framework Convention on Climate Change
United Nations (UN): climate-change litigation and, 30–31; energy rationing and, 6; environmentalism and, 6; global warming and, xiii, 63; Hockey Stick scandal and, 128
Upsala glacier, 223–28
United States: average temperature in, 116fig; carbon dioxide emissions in, 288fig, 289–92; economic growth in, 11; environmental record of, 8–9; fossil fuel consumption in, 116; global cooling and, 116; global warming and, 64; Kyoto Protocol and, xvi, 64, 292–95, 301; temperature in, 115–17
USA Today, 101
U.S. Corporation for Public Broadcasting, 185
U.S. Energy Information Agency, 8
U.S. National Assessment (USNA), 135
U.S. National Assessment on Climate Change, 123
U.S. National Center for Atmosphere Research, 82
U.S. Senate Committee on Environment and Public Works, 6, 77, 84
U.S. Senate Committee on Foreign Relations, 6

V

Virginia, University of, 97
VitalSTATS, 132
von Storch, Hans, 127, 153

W

Wahl, Eugene, 88
Wall Street Journal, 83, 104, 235
Wallstrom, Margaret, 299
Washington Examiner, 269
Washington Post, 46, 71, 78, 94, 162, 171, 199, 235, 238
Watchman, Paul, 31
Waterworld, 147, 184, 188
Watson, Paul, 7
wealth: environmental performance and, 3; as healthier and cleaner, 3, 11, 15–17, 64; redistribution of, 7
weather: climate vs., 64, 173; lies about, 141–65; man and, 76; man's contribution to, 153, 164
Webster, Peter, 151
Weisskopf, Michael, 266
Wendler, Gerd, 132
The West Wing, 171
Whelan, Elizabeth, 9, 15
White House Council on Environmental Quality, 100
WHO. *See* World Health Organization
Wiesel, Elie, 35
Wigley, Thomas, 82, 226, 282
Wildavsky, Aaron, 30
Will, George, 177
Wilson, Edward O., 234
Wirth, Tim, 198
World Energy Outlook (2006), 80
World Glacier Monitoring Service, 223
World Health Organization (WHO), 8, 237, 239
World Rainforest Movement (WRM), 57
World War II, xiv
WorldWatch Institute, 49
WRM. *See* World Rainforest Movement
Wunsch, Carl, 232–33
Wursta, Charles, 17

Z

Zwally, Jay, 147, 149–50